THERMODYNAMICS

BY

ENRICO FERMI

DOVER PUBLICATIONS, INC.

NEW YORK

Published in Canada by General Publishing Company, Ltd., 30 Lesmill Road, Don Mills, Toronto, Ontario.
Published in the United Kingdom by Constable and Company, Ltd., 10 Orange Street, London WC 2.

This Dover edition, first published in 1956, is an unabridged and unaltered republication of the work originally published by Prentice-Hall Company in 1937.

Standard Book Number: 486-60361-X
Library of Congress Catalog Card Number: 57-14599

Manufactured in the United States of America
Dover Publications, Inc.
180 Varick Street
New York, N. Y. 10014

Preface

THIS book originated in a course of lectures held at Columbia University, New York, during the summer session of 1936.

It is an elementary treatise throughout, based entirely on pure thermodynamics; however, it is assumed that the reader is familiar with the fundamental facts of thermometry and calorimetry. Here and there will be found short references to the statistical interpretation of thermodynamics.

As a guide in writing this book, the author used notes of his lectures that were taken by Dr. Lloyd Motz, of Columbia University, who also revised the final manuscript critically. Thanks are due him for his willing and intelligent collaboration.

E. FERMI

Contents

Introduction

THERMODYNAMICS is mainly concerned with the transformations of heat into mechanical work and the opposite transformations of mechanical work into heat.

Only in comparatively recent times have physicists recognized that heat is a form of energy that can be changed into other forms of energy. Formerly, scientists had thought that heat was some sort of fluid whose total amount was invariable, and had simply interpreted the heating of a body and analogous processes as consisting of the transfer of this fluid from one body to another. It is, therefore, noteworthy that on the basis of this heat-fluid theory Carnot was able, in the year 1824, to arrive at a comparatively clear understanding of the limitations involved in the transformation of heat into work, that is, of essentially what is now called the second law of thermodynamics (see Chapter III).

In 1842, only eighteen years later, R. J. Mayer discovered the equivalence of heat and mechanical work, and made the first announcement of the principle of the conservation of energy (the first law of thermodynamics).

We know today that the actual basis for the equivalence of heat and dynamical energy is to be sought in the kinetic interpretation, which reduces all thermal phenomena to the disordered motions of atoms and molecules. From this point of view, the study of heat must be considered as a special branch of mechanics: the mechanics of an ensemble of such an enormous number of particles (atoms or molecules) that the detailed description of the state and the motion loses importance and only average properties of large numbers of particles are to be considered. This branch of mechanics, called *statistical mechanics*, which has been developed mainly through the work of Maxwell, Boltzmann, and Gibbs, has led to a very satisfactory understanding of the fundamental thermodynamical laws.

But the approach in pure thermodynamics is different. Here the fundamental laws are assumed as postulates based on experimental evidence, and conclusions are drawn from them without entering into the kinetic mechanism of the phenomena. This procedure has the advantage of being independent, to a great extent, of the simplifying assumptions that are often made in statistical mechanical considerations. Thus, thermodynamical results are generally highly accurate. On the other hand, it is sometimes rather unsatisfactory to obtain results without being able to see in detail how things really work, so that in many respects it is very often convenient to complete a thermodynamical result with at least a rough kinetic interpretation.

The first and second laws of thermodynamics have their statistical foundation in classical mechanics. In recent years Nernst has added a third law which can be interpreted statistically only in terms of quantum mechanical concepts. The last chapter of this book will concern itself with the consequences of the third law.

CHAPTER I

Thermodynamic Systems

1. The state of a system and its transformations. The state of a system in mechanics is completely specified at a given instant of time if the position and velocity of each mass-point of the system are given. For a system composed of a number N of mass-points, this requires the knowledge of $6N$ variables.

In thermodynamics a different and much simpler concept of the state of a system is introduced. Indeed, to use the dynamical definition of *state* would be inconvenient, because all the systems which are dealt with in thermodynamics contain a very large number of mass-points (the atoms or molecules), so that it would be practically impossible to specify the $6N$ variables. Moreover, it would be unnecessary to do so, because the quantities that are dealt with in thermodynamics are average properties of the system; consequently, a detailed knowledge of the motion of each mass-point would be superfluous.

In order to explain the thermodynamic concept of the state of a system, we shall first discuss a few simple examples.

A system composed of a chemically defined homogeneous fluid. We can make the following measurements on such a system: the temperature t, the volume V, and the pressure p. The temperature can be measured by placing a thermometer in contact with the system for an interval of time sufficient for thermal equilibrium to set in. As is well known, the temperature defined by any special thermometer (for example, a mercury thermometer) depends on the particular properties of the thermometric substance used. For the time being, we shall agree to use the same kind of thermometer for all temperature measurements in order that these may all be comparable.

The geometry of our system is obviously characterized not only by its volume, but also by its shape. However, most thermodynamical properties are largely independent of the shape, and, therefore, the volume is the only geometrical datum that is ordinarily given. It is only in the cases for which the ratio of surface to volume is very large (for example, a finely grained substance) that the surface must also be considered.

For a given amount of the substance contained in the system, the temperature, volume, and pressure are not independent quantities; they are connected by a relationship of the general form:

$$f(p, V, t) = 0, \tag{1}$$

which is called the *equation of state.* Its form depends on the special properties of the substance. Any one of the three variables in the above relationship can be expressed as a function of the other two by solving equation (1) with respect to the given variable. Therefore, the state of the system is completely determined by any two of the three quantities, p, V, and t.

It is very often convenient to represent these two quantities graphically in a rectangular system of co-ordinates. For example, we may use a (V, p) representation, plotting V along the abscissae axis and p along the ordinates axis. A point on the (V, p) plane thus defines a state of the system. The points representing states of equal temperature lie on a curve which is called an *isothermal.*

A system composed of a chemically defined homogeneous solid. In this case, besides the temperature t and volume V, we may introduce the pressures acting in different directions in order to define the state. In most cases, however, the assumption is made that the solid is subjected to an isotropic pressure, so that only one value for the pressure need be considered, as in the case of a fluid.

A system composed of a homogeneous mixture of several chemical compounds. In this case the variables defining the state of the system are not only temperature, volume, and pressure, but also the concentrations of the different chemical compounds composing the mixture.

Nonhomogeneous systems. In order to define the state of a nonhomogeneous system, one must be able to divide it into a number of homogeneous parts. This number may be finite in some cases and infinite in others. The latter possibility, which is only seldom considered in thermodynamics, arises when the properties of the system, or at least of some of its parts, vary continuously from point to point. The state of the system is then defined by giving the mass, the chemical composition, the state of aggregation, the pressure, the volume, and the temperature of each homogeneous part.

It is obvious that these variables are not all independent. Thus, for example, the sum of the amounts of each chemical element present in the different homogeneous parts must be constant and equal to the total amount of that element present in the system. Moreover, the volume, the pressure, and the temperature of each homogeneous part having a given mass and chemical composition are connected by an equation of state.

A system containing moving parts. In almost every system that is dealt with in thermodynamics, one assumes that the different parts of the system either are at rest or are moving so slowly that their kinetic energies may be neglected. If this is not the case, one must also specify the velocities of the various parts of the system in order to define the state of the system completely.

It is evident from what we have said that the knowledge of the thermodynamical state alone is by no means sufficient for the determination of the dynamical state. Studying the thermodynamical state of a homogeneous fluid of given volume at a given temperature (the pressure is then defined by the equation of state), we observe that there is an infinite number of states of molecular motion that correspond to it. With increasing time, the system exists successively in all these dynamical states that correspond to the given thermodynamical state. From this point of view we may say that a thermodynamical state is the ensemble of all the dynamical states through which, as a result of the molecular motion, the system is rapidly passing. This definition of state is rather abstract and not quite unique; therefore, we shall indicate in each particular case what the state variables are.

Particularly important among the thermodynamical states of a system are the *states of equilibrium*. These states have the property of not varying so long as the external conditions remain unchanged. Thus, for instance, a gas enclosed in a container of constant volume is in equilibrium when its pressure is constant throughout and its temperature is equal to that of the environment.

Very often we shall have to consider *transformations* of a system from an initial state to a final state through a continuous succession of intermediate states. If the state of the system can be represented on a (V, p) diagram, such a transformation will be represented by a curve connecting the two points that represent the initial and final states.

A transformation is said to be *reversible* when the successive states of the transformation differ by infinitesimals from *equilibrium states*. A reversible transformation can therefore connect only those initial and final states which are states of equilibrium. A reversible transformation can be realized in practice by changing the external conditions so slowly that the system has time to adjust itself gradually to the altered conditions. For example, we can produce a reversible expansion of a gas by enclosing it in a cylinder with a movable piston and shifting the piston outward very slowly. If we were to shift the piston rapidly, currents would be set up in the expanding gaseous mass, and the intermediate states would no longer be states of equilibrium.

If we transform a system reversibly from an initial state A to a final state B, we can then take the system by means of the reverse transformation from B to A through the same succession of intermediate states but in the reverse order. To do this, we need simply change the conditions of the environment very slowly in a sense opposite to that in the original transformation. Thus, in the case of the gas discussed in the preceding paragraph, we may compress it again to its original volume and bring it back to its initial state by shifting the piston inward very slowly. The compression occurs reversibly, and the gas passes through the same intermediate states as it did during the expansion.

During a transformation, the system can perform positive or negative external *work*; that is, the system can do work on its surroundings or the surroundings can do work on the system. As an example of this, we consider a body enclosed in a cylinder having a movable piston of area S at one end (Figure 1). If p is the pressure of the body against the walls of the cylinder, then pS is the force exerted by the body on the piston. If the piston is shifted an infinitesimal distance dh, an infinitesimal amount of work,

$$dL = pSdh, \tag{2}$$

is performed, since the displacement is parallel to the force. But Sdh is equal to the increase, dV, in volume of the system. Thus, we may write[1]:

$$dL = pdV. \tag{3}$$

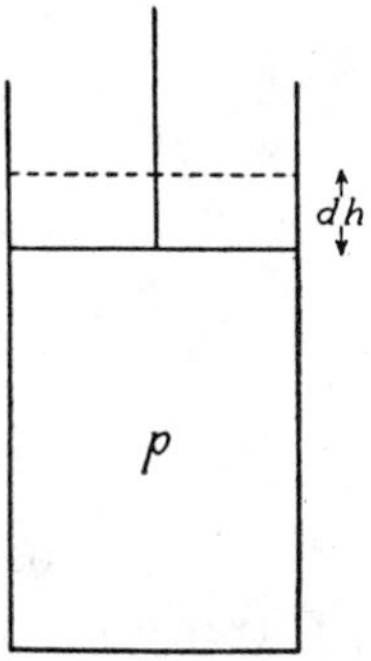

Fig. 1.

[1] It is obvious that (3) is generally valid no matter what the shape of the container may be. Consider a body at the uniform pressure p, enclosed in an irregularly shaped container A (Figure 2). Consider now an infinitesimal transformation of our system during which the walls of the container move from the initial position A to the final position B, thus permitting the body inside the container to expand. Let $d\sigma$ be a surface element of the container, and let dn be the displacement of this element in the direction normal to the surface of the container. The work performed on the surface element $d\sigma$ by the pressure p during the displacement of the container from the situation A to the situation B is obviously $p\ d\sigma\ dn$. The total amount of work performed during the infinitesimal transformation is obtained by integrating the above expression over all the surface σ of the container; since p is a constant, we obtain:

$$dL = p \int d\sigma\ dn.$$

It is now evident from the figure that the variation dV of the volume of the container is given by the surface integral,

$$dV = \int d\sigma\ dn.$$

Comparing these two equations, we obtain (3).

For a finite transformation, the work done by the system is obtained by integrating equation (3):

$$L = \int_A^B p dV, \tag{4}$$

where the integral is taken over the entire transformation.

When the state of the system can be represented on a (V, p) diagram, the work performed during a transformation has a simple geometrical representation. We consider a transformation from an initial state indicated by the point A to a final state indicated by the point B (Figure 3). This transformation will be represented by a curve connecting A and B the shape of which depends on the type of transformation considered. The work done during this transformation is given by the integral

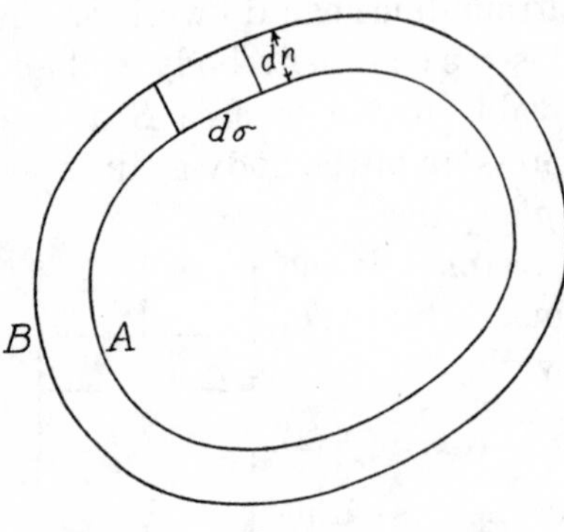

Fig. 2.

$$L = \int_{V_A}^{V_B} p dV, \tag{5}$$

where V_A and V_B are the volumes corresponding to the states A and B. This integral, and hence the work done, can be represented geometrically by the shaded area in the figure.

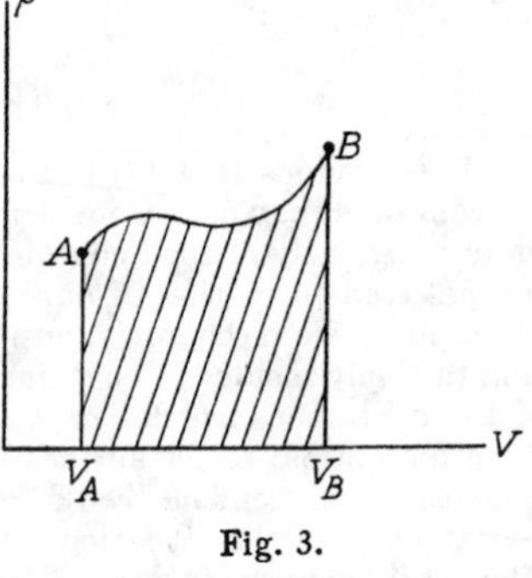

Fig. 3.

Transformations which are especially important are those for which the initial and final states are the same. These are called *cyclical transformations* or *cycles*. A cycle, therefore, is a transformation which brings the system back to its initial state. If the state of the system can be represented on a (V, p) diagram, then a cycle can be represented on this diagram by a closed curve, such as the curve $ABCD$ (Figure 4).

The work, L, performed by the system during the cyclical transformation is given geometrically by the area enclosed by the curve representing the cycle. Let A and C be the points of minimum and maximum abscissa of our cycle, and let their projections on the V-axis be A' and C', respectively. The work performed during the part ABC of the transformation is positive and equal to the area $ABCC'A'A$. The work performed during the rest of the transformation, CDA, is negative and equal in amount to the area $CC'A'ADC$. The total amount of positive work done is equal to the difference between these two areas, and hence is equal to the area bounded by the cycle.

It should be noted that the total work done is positive because we performed the cycle in a clockwise direction. If the same cycle is performed in a counterclockwise direction, the work will again be given by the area bounded by the cycle, but this time it will be negative.

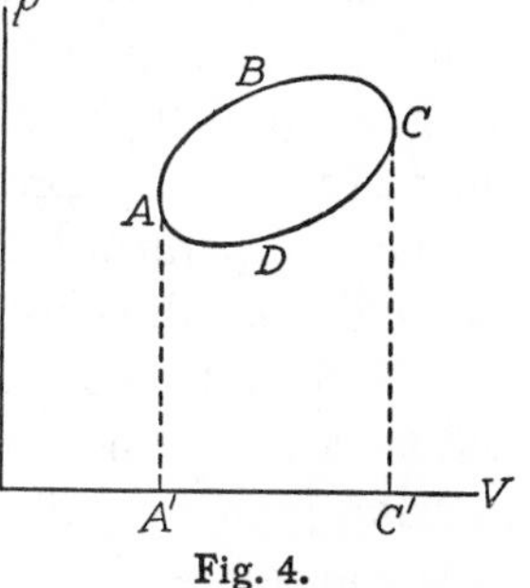

Fig. 4.

A transformation during which the system performs no external work is called an *isochore transformation.* If we assume that the work dL performed during an infinitesimal element of the transformation is given, according to equation (3), by pdV, we find for an isochore transformation $dV = 0$, or, by integration, $V =$ a constant. Thus, an isochore transformation in this case is a transformation at constant volume. This fact justifies the name *isochore.* It should be noticed, however, that the concept of isochore transformation is more general, since it requires that $dL = 0$ for the given transformation, even when the work dL cannot be represented by equation (3).

Transformations during which the pressure or the temperature of the system remains constant are called *isobaric* and *isothermal* transformations, respectively.

2. Ideal or perfect gases. The equation of state of a system composed of a certain quantity of gas occupying a volume V at the temperature t and pressure p can be approximately expressed by a very simple analytical law. We obtain the equation of state of a gas in its simplest form by changing from the empirical scale of temperatures, t, used so far to a new temperature scale T.

We define T provisionally as the temperature indicated by a gas thermometer in which the thermometric gas is kept at a very low constant pressure. T is then taken proportional to the volume occupied by the gas. It is well known that the readings of different gas thermometers under these conditions are largely independent of the nature of the thermometric gas, provided that this gas is far enough from condensation. We shall see later, however (section 9), that it is possible to define this same scale of temperatures T by general thermodynamic considerations quite independently of the special properties of gases.

The temperature T is called the *absolute temperature.* Its unit is usually chosen in such a way that the temperature difference between the boiling and the freezing points of water at one atmosphere of pressure is equal to 100. The freezing point of water corresponds then, as is well known, to the absolute temperature 273.1.

The equation of state of a system composed of m grams of a gas whose molecular weight is M is given approximately by:

$$pV = \frac{m}{M} RT. \tag{6}$$

R is a universal constant (that is, it has the same value for all gases: $R = 8.314 \times 10^7$ erg/degrees, or (see section 3) $R = 1.986$ cal/degrees). Equation (6) is called *the equation of state of an ideal or a perfect gas*; it includes the laws of Boyle, Gay-Lussac, and Avogadro.

No real gas obeys equation (6) exactly. An ideal substance that obeys equation (6) exactly is called an ideal or a perfect gas.

For a gram-molecule (or mole) of a gas (that is, for a number of grams of a gas equal numerically to its molecular weight), we have $m = M$, so that (6) reduces to:

$$pV = RT. \tag{7}$$

From (6) or (7) we can obtain the density ρ of the gas in terms of the pressure and the temperature:

$$\rho = \frac{m}{V} = \frac{Mp}{RT}. \tag{8}$$

For an isothermal transformation of an ideal gas (transformation at constant temperature), we have:

$$pV = \textit{constant}.$$

On the (V, p) diagram the isothermal transformations of an ideal gas are thus represented by equilateral hyperbolas having the V- and p-axes as asymptotes.

We can easily calculate the work performed by the gas during an isothermal expansion from an initial volume V_1 to a final volume V_2. This is given (making use of (5) and (6)) by:

$$\begin{aligned} L = \int_{V_1}^{V_2} p dV &= \frac{m}{M} RT \int_{V_1}^{V_2} \frac{dV}{V} \\ &= \frac{m}{M} RT \log \frac{V_2}{V_1} \\ &= \frac{m}{M} RT \log \frac{p_1}{p_2}, \end{aligned} \tag{9}$$

where p_1 and p_2 are the initial and final pressures, respectively. For one mole of gas, we have:

$$L = RT \log \frac{V_2}{V_1} = RT \log \frac{p_1}{p_2}. \tag{10}$$

A mixture of several gases is governed by laws very similar to those which are obeyed by a chemically homogeneous gas. We shall call the *partial pressure* of a component of a mixture of gases the pressure which this component would exert if it alone filled the volume occupied by the mixture at the same temperature as that of the mixture. We can now state Dalton's law for gas mixtures in the following form:

The pressure exerted by a mixture of gases is equal to the sum of the partial pressures of all the components present in the mixture.

This law is only approximately obeyed by real gases, but it is assumed to hold exactly for ideal gases.

Problems

1. Calculate the work performed by a body expanding from an initial volume of 3.12 liters to a final volume of 4.01 liters at the pressure of 2.34 atmospheres.

2. Calculate the pressure of 30 grams of hydrogen inside a container of 1 cubic meter at the temperature of 18°C.

3. Calculate the density and specific volume of nitrogen at the temperature of 0°C.

4. Calculate the work performed by 10 grams of oxygen expanding isothermally at 20°C from 1 to .3 atmospheres of pressure.

CHAPTER II

The First Law of Thermodynamics

3. The statement of the first law of thermodynamics. The first law of thermodynamics is essentially the statement of the principle of the conservation of energy for thermodynamical systems. As such, it may be expressed by stating that the variation in energy of a system during any transformation is equal to the amount of energy that the system receives from its environment. In order to give a precise meaning to this statement, it is necessary to define the phrases "energy of the system" and "energy that the system receives from its environment during a transformation."

In purely mechanical conservative systems, the energy is equal to the sum of the potential and the kinetic energies, and hence is a function of the dynamical state of the system; because to know the dynamical state of the system is equivalent to knowing the positions and velocities of all the mass-points contained in the system. If no external forces are acting on the system, the energy remains constant. Thus, if A and B are two successive states of an isolated system, and U_A and U_B are the corresponding energies, then

$$U_A = U_B.$$

When external forces act on the system, U_A need no longer be equal to U_B. If $-L$ is the work performed by the external forces during a transformation from the initial state A to the final state B ($+L$ is the work performed by the system), then the dynamical principle of the conservation of energy takes the form:

$$U_B - U_A = -L. \tag{11}$$

From this equation it follows that the work, L, performed during the transformation depends only on the extreme states A and B of the transformation and not on the particular way in which the transformation from A to B is performed.

Let us assume now that we do not know the laws of interaction among the various mass-points of our dynamical system. Then we cannot calculate the energy of the system when it is in a given dynamical state. By making use of equation (11), however, we can nevertheless obtain an empirical definition of the energy of our system in the following way:

We consider an arbitrarily chosen state O of our system and, by definition, take its energy to be zero:

$$U_O = 0. \tag{12}$$

We shall henceforth refer to this state as the *standard* state of our system. Consider now any other state A; by applying suitable external forces to our system, we can transform it from the standard state (in which we assume it to be initially) to the state A. Let L_A be the work performed by the system during this transformation ($-L_A$ is, as before, the work performed by the external forces on the system). Applying (11) to this transformation, and remembering (12), we find that

$$U_A = -L_A. \tag{13}$$

This equation can be used as the empirical definition of the energy U_A of our system in the state A.

It is obviously necessary, if definition (13) is to have a meaning, that the work L_A depend only on the states O and A and not on the special way in which the transformation from O to A is performed. We have already noticed that this property follows from (11). If one found experimentally that this property did not hold, it would mean either that energy is not conserved in our system, or that, besides mechanical work, other means of transfer of energy must be taken into account.

We shall assume for the present that the work performed by our mechanical system during any transformation depends only on the initial and final states of the transformation, so that we can use (13) as the definition of the energy.

We can immediately obtain (11) from (13) as follows: A transformation between any two states A and B can always be performed as a succession of two transformations: first a transformation from A to the standard state O, and then a transformation from O to B. Since the system performs the amounts of work $-L_A$ and $+L_B$ during these two transformations, the total amount of work performed during the transformation from A to B (which is independent of the particular way in which the transformation is performed) is:

$$L = -L_A + L_B.$$

From (13) and the analogous equation,

$$U_B = -L_B,$$

we obtain now:

$$U_B - U_A = -L,$$

which is identical with (11).

We notice, finally, that the definition (13) of the energy is not quite unique, since it depends on the particular choice of the standard state O. If instead of O we had chosen a different standard state, O', we should have obtained a different value, U'_A, for the energy of the state A. It can be easily shown, however, that U'_A and U_A differ only by an additive constant. Indeed, the transformation from O' to A can be put equal to the sum of two transformations: one going from O' to O and the other going from O to A. The work L'_A performed by the system in passing from O' to A is thus equal to:

$$L'_A = L_{O'O} + L_A,$$

where $L_{O'O}$ is the work performed by the system in going from O' to O. We have now:

$$U_A = -L_A; \qquad U'_A = -L'_A,$$

so that

$$U_A - U'_A = L_{O'O},$$

which shows that the values of the energy based on the two definitions differ only by the constant $L_{O'O}$.

This indeterminate additive constant which appears in the definition of the energy is, as is well known, an essential feature of the concept of energy. Since, however, only differences of energy are considered in practice, the additive constant does not appear in the final results.

The only assumption underlying the above empirical definition of the energy is that the total amount of work performed by the system during any transformation depends only on the initial and final states of the transformation. We have already noticed that if this assumption is contradicted by experiment, and if we still do not wish to discard the principle of the conservation of energy, then we must admit the existence of other methods, besides mechanical work, by means of which energy can be exchanged between the system and its environment.

Let us take, for example, a system composed of a quantity of water. We consider two states A and B of this system at atmospheric pressure; let the temperatures of the system in these two states be t_A and t_B, respectively, with $t_A < t_B$. We can take our system from A to B in two different ways.

First way: We heat the water by placing it over a flame and raise its temperature from the initial value t_A to the final value t_B. The external work performed by the system during this transformation is practically zero. It would be exactly zero if the change in temperature were not accompanied by a change in volume of the water. Actually, however, the volume of the water changes slightly during the transformation, so that a small amount of work is performed (see equation (3)). We shall neglect this small amount of work in our considerations.

Second way: We raise the temperature of the water from t_A to t_B by heating it by means of friction. To this end, we immerse a small set of paddles attached to a central axle in the water, and churn the water by rotating the paddles. We observe that the temperature of the water increases continuously as long as the paddles continue to rotate. Since the water offers resistance to the motion of the paddles, however, we must perform mechanical work in order to keep the paddles moving until the final temperature t_B is reached. Corresponding to this considerable amount of positive work performed by the paddles on the water, there is an equal amount of negative work performed by the water in resisting the motion of the paddles.

We thus see that the work performed by the system in going from the state A to the state B depends on whether we go by means of the first way or by means of the second way.

If we assume that the principle of the conservation of energy holds for our system, then we must admit that the energy that is transmitted to the water in the form of the mechanical work of the rotating paddles in the second way is transmitted to the water in the first way in a nonmechanical form called *heat*. We are thus led to the fact that heat and mechanical work are equivalent; they are two different aspects of the same thing, namely, energy. In what follows we shall group under the name of work electrical and magnetic work as well as mechanical work. The first two types of work, however, are only seldom considered in thermodynamics.

In order to express in a more precise form the fact that heat and work are equivalent, we proceed as follows.

We first enclose our system in a container with non-heat-conducting walls in order to prevent exchange of heat with the environment.[1] We assume, however, that work can be exchanged between the system and its environment (for example, by enclosing the system in a cylinder with non-conducting walls but with a movable piston at one end). The exchange of energy between the inside and the outside of the container can now occur only in the form of work, and from the principle of the conservation of energy it follows that the amount of work performed by the system during any transformation depends only on the initial and the final states of the transformation.[2]

We can now use the empirical definition (13) of the energy and define the energy U as a function of the state of the system only.[3] Denoting by $\Delta U = U_B - U_A$ the variation in the energy of our system that occurs during a transformation from the state A to the state B, we can write equation (11), which is applicable to our thermally insulated system, in the form:

$$\Delta U + L = 0. \tag{14}$$

If our system is not thermally insulated, the left-hand side of (14) will in general be different from zero because there can then take place an exchange of energy in the form of

---

[1] We need only mention here that no perfect thermal insulators exist. Thermal insulation can be obtained approximately, however, by means of the well-known methods of Calorimetry.

[2] It would be formally more exact, although rather abstract, to state the content of the preceding sentences as follows:

Experiments show that there exist certain substances called *thermal insulators* having the following properties: when a system is completely enclosed in a thermal insulator in such a way that work can be exchanged between the inside and the outside, the amount of work performed by the system during a given transformation depends only on the initial and final states of the transformation.

[3] It should be noticed here that if definition (13) of the energy of a state A of our system is to be applicable, it must be possible to transform the system from the standard state O to the state A while the system is thermally insulated. We shall show later (see section 13) that such a transformation is not always possible without an exchange of heat. In such cases, however, the opposite transformation $A \rightarrow O$ can always be performed. The work performed by the system during this reverse transformation is $-L_A$; we can therefore apply (13) to such cases also.

heat. We shall therefore replace (14) by the more general equation:

$$\Delta U + L = Q, \tag{15}$$

where Q is equal to zero for transformations performed on thermally insulated systems and otherwise, in general, is different from zero.

Q can be interpreted physically as the amount of energy that is received by the system in forms other than work. This follows immediately from the fact that the variation in energy, ΔU, of the system must be equal to the total amount of energy received by the system from its environment. But from (15)

$$\Delta U = -L + Q,$$

and $-L$ is the energy received in the form of work. Hence, Q stands for the energy received in all other forms.

By definition, we shall now call Q the amount of heat received by the system during the transformation.

For a cyclic transformation, equation (15) takes on a very simple form. Since the initial and final states of a cycle are the same, the variation in energy is zero: $\Delta U = 0$. Thus, (15) becomes:

$$L = Q. \tag{16}$$

That is, the work performed by a system during a cyclic transformation is equal to the heat absorbed by the system.

It is important at this point to establish the connection between this abstract definition of heat and its elementary calorimetric definition. The calorimetric unit of heat, the *calorie*, is defined as the quantity of heat required to raise the temperature of one gram of water at atmospheric pressure from 14°C to 15°C. Thus, to raise the temperature of m grams of water from 14°C to 15°C at atmospheric pressure, we require m calories of heat. Let Δu_c denote the variation in energy of one gram of water, and l_c the work done as a result of its expansion when its temperature is raised from 14°C to 15°C at atmospheric pressure. For m grams of water, the variation in energy and the work done are:

$$\Delta U_c = m\Delta u_c; \qquad L_c = ml_c. \tag{17}$$

We now consider a system S which undergoes a transformation. In order to measure the heat exchanged between the system and the surrounding bodies, we place the system in contact with a calorimeter containing m grams of water, initially at 14°C. We choose the mass of the water in such a way that after the transformation has been completed, the temperature of the water is 15°C.

Since an ideal calorimeter is perfectly insulated thermally, the complex system composed of the system S and the calorimetric water is thermally insulated during the transformation. We may therefore apply equation (14) to this transformation. The total variation in energy is equal to the sum:

$$\Delta U = \Delta U_S + \Delta U_c,$$

where ΔU_S is the variation in energy of the system S, and ΔU_c is the variation in energy of the calorimetric water. Similarly, for the total work done, we have:

$$L = L_S + L_c.$$

From (14) we have, then,

$$\Delta U_S + \Delta U_c + L_S + L_c = 0;$$

or, by (17),

$$\begin{aligned} \Delta U_S + L_S &= -(\Delta U_c + L_c) \\ &= -m(\Delta u_c + l_c). \end{aligned}$$

But from the definition (15), $\Delta U_S + L_S$ is the amount of heat Q_S received by the system S. Thus, we have:

$$Q_S = -m(\Delta u_c + l_c). \tag{18}$$

We see from this that the amount of heat is proportional to m.

On the other hand, in calorimetry the fact that m grams of calorimetric water have been heated from 14°C to 15°C means that m calories of heat have been transferred from the system S to the calorimeter; that is, that the system S has received $-m$ calories, or that Q_S, expressed in calories, is equal to $-m$. We see also, by comparison with (18), that the amount of heat, as given by the definition (15), is proportional to the amount when it is expressed in calories; the constant of proportionality is $(\Delta u_c + l_c)$.

According to (15), heat is measured in energy units (ergs). The constant ratio between ergs and calories has been measured by many investigators, who have found that

$$1 \text{ calorie} = 4.185 \times 10^7 \text{ ergs.} \tag{19}$$

In what follows we shall generally express heat measurements in energy units.

Equation (15), which is a precise formulation of the equivalence of heat and work, expresses the *first law of thermodynamics*.

4. The application of the first law to systems whose states can be represented on a (V, p) diagram. We shall now apply the first law of thermodynamics to a system, such as a homogeneous fluid, whose state can be defined in terms of any two of the three variables V, p, and T. Any function of the state of the system, as, for example, its energy, U, will then be a function of the two variables which have been chosen to represent the state.

In order to avoid any misunderstanding as to which are the independent variables when it is necessary to differentiate partially, we shall enclose the partial derivative symbol in a parenthesis and place the variable that is to be held constant in the partial differentiation at the foot of the parenthesis. Thus, $\left(\frac{\partial U}{\partial T}\right)_V$ means the derivative of

U with respect to T, keeping V constant, when T and V are taken as the independent variables. Notice that the above expression is in general different from $\left(\frac{\partial U}{\partial T}\right)_p$, because in the first case the volume is kept constant while in the second case the pressure is kept constant.

We now consider an infinitesimal transformation of our system, that is, a transformation for which the independent variables change only by infinitesimal amounts. We apply to this transformation the first law of thermodynamics as expressed by equation (15). Instead of ΔU, L, and Q, we must now write dU, dL, and dQ, in order to point out the infinitesimal nature of these quantities. We obtain, then,

$$dU + dL = dQ. \tag{20}$$

Since for our system, dL is given by (3), we have:

$$dU + pdV = dQ. \tag{21}$$

If we choose T and V as our independent variables, U becomes a function of these variables, so that:

$$dU = \left(\frac{\partial U}{\partial T}\right)_V dT + \left(\frac{\partial U}{\partial V}\right)_T dV,$$

and (21) becomes:

$$\left(\frac{\partial U}{\partial T}\right)_V dT + \left[\left(\frac{\partial U}{\partial V}\right)_T + p\right] dV = dQ. \tag{22}$$

Similarly, taking T and p as independent variables, we have:

$$\left[\left(\frac{\partial U}{\partial T}\right)_p + p\left(\frac{\partial V}{\partial T}\right)_p\right] dT + \left[\left(\frac{\partial U}{\partial p}\right)_T + p\left(\frac{\partial V}{\partial p}\right)_T\right] dp = dQ. \tag{23}$$

Finally, taking V and p as independent variables, we obtain:

$$\left(\frac{\partial U}{\partial p}\right)_V dp + \left[\left(\frac{\partial U}{\partial V}\right)_p + p\right] dV = dQ. \tag{24}$$

The *thermal capacity* of a body is, by definition, the ratio, dQ/dT, of the infinitesimal amount of heat dQ absorbed by the body to the infinitesimal increase in temperature dT

produced by this heat. In general, the thermal capacity of a body will be different according as to whether the body is heated at constant volume or at constant pressure. Let C_V and C_p be the thermal capacities at constant volume and at constant pressure, respectively.

A simple expression for C_V can be obtained from (22). For an infinitesimal transformation at constant volume, $dV = 0$; hence,

$$C_V = \left(\frac{dQ}{dT}\right)_V = \left(\frac{\partial U}{\partial T}\right)_V. \tag{25}$$

Similarly, using (23), we obtain the following expression for C_p :

$$C_p = \left(\frac{dQ}{dT}\right)_p = \left(\frac{\partial U}{\partial T}\right)_p + p\left(\frac{\partial V}{\partial T}\right)_p. \tag{26}$$

The second term on the right-hand side represents the effect on the thermal capacity of the work performed during the expansion. An analogous term is not present in (25), because in that case the volume is kept constant so that no expansion occurs.

The thermal capacity of one gram of a substance is called the *specific heat* of that substance; and the thermal capacity of one mole is called the *molecular heat.* The specific and molecular heats at constant volume and at constant pressure are given by the formulae (25) and (26) if, instead of taking an arbitrary amount of substance, we take one gram or one mole of the substance, respectively.

5. The application of the first law to gases. In the case of a gas, we can express the dependence of the energy on the state variables explicitly. We choose T and V as the independent variables, and prove first that the energy is a function of the temperature T only and does not depend on the volume V. This, like many other properties of gases, is only approximately true for real gases and is assumed to hold exactly for ideal gases. In section 14 we shall deduce from the second law of thermodynamics the result that the energy of any body which obeys the equation of state, (7), of an ideal gas must be independent of the volume V. At this point, however, we shall give an experimental proof of this proposition for a gas; the experiment was performed by Joule.

Into a calorimeter Joule placed a container having two chambers, A and B, connected by a tube (Figure 5). He filled the chamber A with a gas and evacuated B, the two chambers having first been shut off from each other by a stopcock in the connecting tube. After thermal equilibrium had set in, as indicated by a thermometer placed within the calorimeter, Joule opened the stopcock, thus permitting the gas to flow from A into B until the pressure everywhere in the container was the same. He then observed that there was only a very slight change in the reading of the thermometer. This meant that there had been practically no transfer of heat from the calorimeter to the chamber or vice versa. It is assumed that if this experiment could be performed with an ideal gas, there would be no temperature change at all.

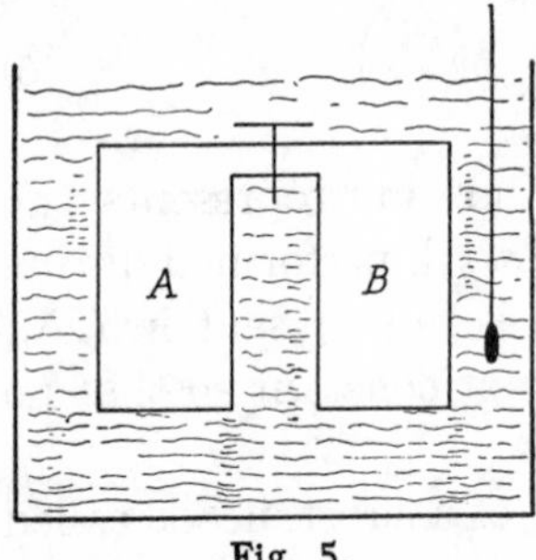

Fig. 5.

We now apply the first law to the above transformation. Since $Q = 0$, we have from equation (15) for the system composed of the two chambers and the enclosed gas:

$$\Delta U + L = 0,$$

where L is the work performed by the system and ΔU is the variation in energy of the system. Since the volumes of the two chambers A and B composing our system do not change during the experiment, our system can perform no external work, that is, $L = 0$. Therefore,

$$\Delta U = 0;$$

the energy of the system, and, hence, the energy of the gas, do not change.

Let us now consider the process as a whole. Initially the gas occupied the volume A, and at the end of the process it filled the two chambers A and B; that is, the transformation resulted in a change in volume of the gas. The experiment showed, however, that there was no resultant change in the temperature of the gas. Since there was no variation in energy during the process, we must conclude that a variation in volume at constant temperature produces no variation in energy. In other words, *the energy of an ideal gas is a function of the temperature only and not a function of the volume.* We may therefore write for the energy of an ideal gas:

$$U = U(T). \tag{27}$$

In order to determine the form of this function, we make use of the experimental result that the specific heat at constant volume of a gas depends only slightly on the temperature; we shall assume that for an ideal gas the specific heat is exactly constant. In this section we shall always refer to one mole of gas; C_V and C_p will therefore denote the molecular heats at constant volume and at constant pressure, respectively.

Since U depends only on T, it is not necessary to specify that the volume is to be kept constant in the derivative in (25); so that, for an ideal gas, we may write:

$$C_V = \frac{dU}{dT}. \tag{28}$$

Since C_V is assumed to be constant, we can integrate at once, and we get:

$$U = C_V T + W, \tag{29}$$

where W is a constant of integration which represents the energy left in the gas at absolute zero temperature.[4]

[4] This additive constant affects the final results of the calculations only when chemical transformations or changes of the states of aggregation of the substances are involved. (See, for example, Chapter VI.) In all other cases, one may place the additive constant equal to zero.

For an ideal gas, equation (21), which expresses the first law of thermodynamics for infinitesimal transformations, takes on the form:

$$C_V dT + p dV = dQ. \tag{30}$$

Differentiating the characteristic equation (7) for one mole of an ideal gas, we obtain:

$$p dV + V dp = R dT. \tag{31}$$

Substituting this in (30), we find:

$$(C_V + R) dT - V dp = dQ. \tag{32}$$

Since $dp = 0$ for a transformation at constant pressure, this equation gives us:

$$C_p = \left(\frac{dQ}{dT}\right)_p = C_V + R. \tag{33}$$

That is, the difference between the molecular heats of a gas at constant pressure and at constant volume is equal to the gas constant R.

The same result may also be obtained from (26), (29), and (7). Indeed, for an ideal gas we have from (29) and (7):

$$\left(\frac{\partial U}{\partial T}\right)_p = \frac{dU}{dT} = C_V; \qquad \left(\frac{\partial V}{\partial T}\right)_p = \left(\frac{\partial}{\partial T}\frac{RT}{p}\right)_p = \frac{R}{p}.$$

Substituting these expressions in (26), we again obtain (33).

It can be shown by an application of kinetic theory that:

$$\begin{aligned} C_V &= \tfrac{3}{2} R \text{ for a monatomic gas; and} \\ C_V &= \tfrac{5}{2} R \text{ for a diatomic gas.} \end{aligned} \tag{34}$$

Assuming these values, which are in good agreement with experiment, we deduce from (33) that:

$$\begin{aligned} C_p &= \tfrac{5}{2} R \text{ for a monatomic gas; and} \\ C_p &= \tfrac{7}{2} R \text{ for a diatomic gas.} \end{aligned} \tag{35}$$

If we place

$$K = \frac{C_p}{C_V} = \frac{C_V + R}{C_V} = 1 + \frac{R}{C_V}, \tag{36}$$

we also obtain:

$$K = \tfrac{5}{3} \text{ for a monatomic gas; and}$$
$$K = \tfrac{7}{5} \text{ for a diatomic gas.} \tag{37}$$

6. Adiabatic transformations of a gas. A transformation of a thermodynamical system is said to be *adiabatic* if it is reversible and if the system is thermally insulated so that no heat can be exchanged between it and its environment during the transformation.

We can expand or compress a gas adiabatically by enclosing it in a cylinder with non-heat-conducting walls and piston, and shifting the piston outward or inward very slowly. If we permit a gas to expand adiabatically, it does external work, so that L in equation (15) is positive. Since the gas is thermally insulated, $Q = 0$, and, hence, ΔU must be negative. That is, the energy of a gas decreases during an adiabatic expansion. Since the energy is related to the temperature through equation (29), a decrease in energy means a decrease in the temperature of the gas also.

In order to obtain a quantitative relationship between the change in temperature and the change in volume resulting from an adiabatic expansion of a gas, we observe that, since $dQ = 0$, equation (30) becomes:

$$C_V dT + p dV = 0.$$

Using the equation of state, $pV = RT$, we can eliminate p from the above equation and obtain:

$$C_V dT + \frac{RT}{V} dV = 0,$$

or

$$\frac{dT}{T} + \frac{R}{C_V}\frac{dV}{V} = 0.$$

Integration yields:

$$\log T + \frac{R}{C_V} \log V = \text{constant}.$$

Changing from logarithims to numbers, we get:

$$TV^{\frac{R}{C_V}} = \text{constant.}$$

Making use of (36), we can write the preceding equation in the form:

$$TV^{K-1} = \text{constant.} \tag{38}$$

This equation tells us quantitatively how an adiabatic change in the volume of an ideal gas determines the change in its temperature. If, for example, we expand a diatomic gas adiabatically to twice its initial volume, we find from (38) (assuming, according to (37), that $K = \frac{7}{5}$) that the temperature is reduced in the ratio $1:2^{0.4} = 1:1.32$.

Using the equation of state, $pV = RT$, we can put equation (38) of an adiabatic transformation in the following forms:

$$pV^{K} = \text{constant.} \tag{39}$$

$$\frac{T}{p^{\frac{K-1}{K}}} = \text{constant.} \tag{40}$$

Equation (39) is to be compared with the equation,

$$pV = \text{constant,}$$

of an isothermal transformation. On the (V, p) diagram, the isothermals are a family of equilateral hyperbolae; the adiabatic lines represented by equation (39), are qualitatively similar to hyperbolae, but they are steeper because $K > 1$.

Isothermal and adiabatic curves are represented in Figure 6, the former by the solid lines and the latter by the dotted lines.

An interesting and simple application of the adiabatic expansion of a gas is the calculation of the dependence of the temperature of the atmosphere on the height above sea level. The principal reason for this variation of temperature with height above sea level is that there are convection currents in the troposphere which continually transport air from the lower regions to the higher ones and from the higher regions to the lower ones. When air from sea level rises to the upper regions of lower pressure, it expands. Since air is a poor conductor of heat, very little heat is transferred to or from the expanding air, so that we may consider the expansion as taking place adiabatically. Consequently, the temperature of the rising air decreases. On the other hand, air from the upper regions of the atmosphere suffers an adiabatic compression, and hence an increase in temperature, when it sinks to low regions.

In order to calculate the change in temperature, we consider a column of air of unit cross section, and focus our attention on a slab, of height dh, having its lower face at a distance h above sea level. If p is the pressure on the lower face, then the pressure on the upper face will be $p + dp$, where dp is the change in pressure which is due to the weight of the air contained in the slab. If g is the acceleration of gravity and ρ is the density of the air, then the weight of the air in the slab is $\rho g dh$. Since an increase in height is followed by a decrease in pressure, we have:

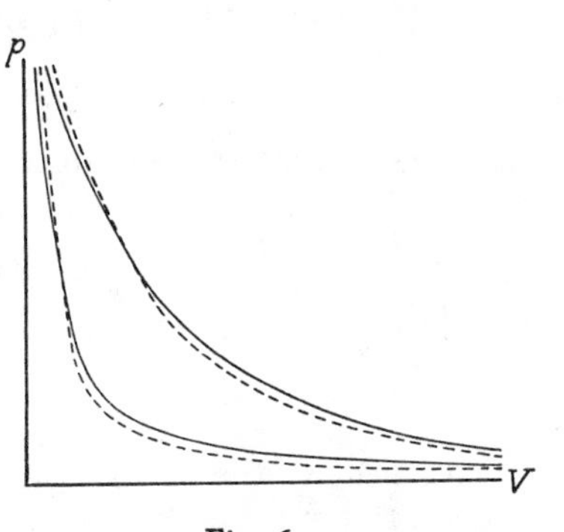

Fig. 6.

$$dp = -\rho g dh; \tag{41}$$

or, remembering (8),

$$dp = -\frac{gM}{R}\frac{p}{T}dh,$$

where M is the average molecular weight of air; $M = 28.88$. The logarithmic derivative of (40) gives us:

$$\frac{dT}{T} = \frac{K-1}{K}\frac{dp}{p}.$$

This, together with the previous equation, gives:

$$\frac{dT}{dh} = -\frac{K-1}{K}\frac{gM}{R}. \tag{42}$$

Assuming

$$K = \tfrac{7}{5}; \quad g = 980.665; \quad M = 28.88; \quad R = 8.214 \times 10^7,$$

we obtain:

$$\frac{dT}{dh} = -9.8 \times 10^{-5} \text{ degrees/cm.}$$

$$= -9.8 \text{ degrees/kilometer.}$$

This value is actually somewhat larger than the observed average decrease of temperature with altitude. The difference is mainly owing to our having neglected the effect of condensation of water vapor in the expanding masses of air.

Problems

1. Calculate the energy variation of a system which performs 3.4×10^8 ergs of work and absorbs 32 calories of heat.

2. How many calories are absorbed by 3 moles of an ideal gas expanding isothermally from the initial pressure of 5 atmospheres to the final pressure of 3 atmospheres, at the temperature of 0°C?

3. One mole of a diatomic ideal gas performs a transformation from an initial state for which temperature and volume are, respectively, 291°K and 21,000 cc. to a final state in which temperature and volume are 305°K and 12,700 cc. The transformation is represented on the (V, p) diagram by a straight line. To find the work performed and the heat absorbed by the system.

4. A diatomic gas expands adiabatically to a volume 1.35 times larger than the initial volume. The initial temperature is 18°C. Find the final temperature.

CHAPTER III

The Second Law of Thermodynamics

7. The statement of the second law of thermodynamics. The first law of thermodynamics arose as the result of the impossibility of constructing a machine which could create energy. The first law, however, places no limitations on the possibility of transforming energy from one form into another. Thus, for instance, on the basis of the first law alone, the possibility of transforming heat into work or work into heat always exists provided the total amount of heat is equivalent to the total amount of work.

This is certainly true for the transformation of work into heat: A body, no matter what its temperature may be, can always be heated by friction, receiving an amount of energy in the form of heat exactly equal to the work done. Similarly, electrical energy can always be transformed into heat by passing an electric current through a resistance. There are very definite limitations, however, to the possibility of transforming heat into work. If this were not the case, it would be possible to construct a machine which could, by cooling the surrounding bodies, transform heat, taken from its environment, into work.

Since the supply of thermal energy contained in the soil, the water, and the atmosphere is practically unlimited, such a machine would, to all practical purposes, be equivalent to a *perpetuum mobile*, and is therefore called a *perpetuum mobile* of the second kind.

The second law of thermodynamics rules out the possibility of constructing a *perpetuum mobile* of the second kind. In order to give a precise statement of this law, we shall define what is meant by a source of heat of a given temperature.

A body which is at the temperature t throughout and is

conditioned in such a way that it can exchange heat but no work with its surroundings is called a source of heat of temperature t. As examples of this, we may consider bodies enclosed in rigid containers or bodies which undergo negligible variations of volume. A mass of water which is at the temperature t throughout may be taken as a source of heat, since its volume remains practically constant.

We can now state the second law of thermodynamics in the following form:

A transformation whose only final result is to transform into work heat extracted from a source which is at the same temperature throughout is impossible.[1] (Postulate of Lord Kelvin.)

The experimental evidence in support of this law consists mainly in the failure of all efforts that have been made to construct a *perpetuum mobile* of the second kind.

The second law can also be expressed as follows:

A transformation whose only final result is to transfer heat from a body at a given temperature to a body at a higher temperature is impossible. (Postulate of Clausius.)

Until now we have made use only of an empirical temperature scale. In order to give a precise meaning to the postulate of Clausius, we must first define what we mean

---

[1] An essential part of Lord Kelvin's postulate is that the transformation of the heat into work be the *only* final result of the process. Indeed, it is not impossible to transform into work heat taken from a source all at one temperature provided some other change in the state of the system is present at the end of the process.

Consider, for example, the isothermal expansion of an ideal gas that is kept in thermal contact with a source of heat at the temperature T. Since the energy of the gas depends only on the temperature, and the temperature does not change during the process, we must have $\Delta U = 0$. From the first law, equation (15), we obtain, then, $L = Q$. That is, the work, L, performed by the expanding gas is equal to the heat Q which it absorbs from the source. There is thus a complete transformation of heat, Q, into work L. This, however, is not a contradiction of Kelvin's postulate, since the transformation of Q into L is not the only final result of the process. At the end of the process, the gas occupies a volume larger than it did at the beginning.

when we say that one body is at a higher temperature than another body. If we bring two bodies at different temperatures into thermal contact, heat flows spontaneously by conduction from one of these bodies to the other. By definition, we shall now say that the body away from which heat flows is at a higher temperature than the other body. With this understanding, we can now state the postulate of Clausius as follows:

If heat flows by conduction from a body ***A*** *to another body* ***B****, then a transformation whose only final result is to transfer heat from* ***B*** *to* ***A*** *is impossible.*

We must now prove the equivalence of the Clausius and the Kelvin postulates. To do this we shall prove that if the Clausius postulate were not valid, the Kelvin postulate would not be valid, and vice versa.

Let us first suppose that Kelvin's postulate were not valid. Then we could perform a transformation whose only final result would be to transform completely into work a definite amount of heat taken from a single source at the temperature t_1. By means of friction we could then transform this work into heat again and with this heat raise the temperature of a given body, regardless of what its initial temperature, t_2, might have been. In particular, we could take t_2 to be higher than t_1. Thus, the only final result of this process would be the transfer of heat from one body (the source at the temperature t_1) to another body at a higher temperature, t_2. This would be a violation of the Clausius postulate.

The second part of the proof of the equivalence of the two postulates requires first a discussion of the possibilities of transforming heat into work. We give this discussion in the next section.

8. The Carnot cycle. Since, according to Kelvin's postulate, it is impossible to transform into work heat taken from a source at a uniform temperature by a transformation that leaves no other change in the systems involved in it, we need at least two sources at different temperatures t_1 and t_2 in order to perform such a transformation. If we have two such sources, we can transform heat into work by the following process, which is called a *Carnot cycle.*

Consider a fluid whose state can be represented on a (V, p) diagram, and consider two adiabatics and two isothermals corresponding to the temperatures t_1 and t_2. These four curves intersect each other in the four points A, B, C, and D, as shown in Figure 7. Let AB and CD be the two isothermal lines having the temperatures t_2 and t_1, respectively. AC and BD are the two adiabatic lines. The reversible cyclic transformation $ABDCA$ is called a *Carnot cycle.*

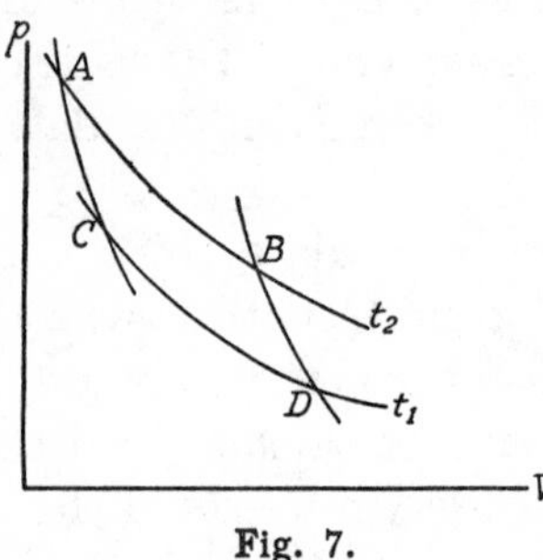

Fig. 7.

The following example will illustrate how a Carnot cycle can actually be performed. We enclose our fluid in a cylindrical container which has nonconducting lateral walls and a nonconducting piston at one end, so that heat can leave or enter the cylinder only through the other end (the base of the cylinder), which we take to be heat-conducting. Let t_1 and t_2 be two sources of heat that are so large that their temperatures remain sensibly unaltered when any finite amounts of heat are added to or subtracted from them. Let t_2 be larger than t_1.

We assume that initially the volume and the pressure of the fluid inside the cylinder are V_A and p_A, respectively, corresponding to the point A in Figure 7. Since this point lies on the isothermal corresponding to the temperature t_2, the temperature of the fluid is equal to t_2 initially. If, therefore, we place the cylinder on the source t_2, no transfer of heat will occur (Figure 8, A). Keeping the cylinder on the source t_2, we raise the piston very slowly and thus increase the volume reversibly until it has reached the value

V_B (Figure 8, B). This part of the transformation is represented by the segment AB of the isothermal t_2. The state of our system is now represented by the point B in Figure 7.

We now place the cylinder on a thermal insulator and increase the volume very slowly until it has reached the value V_D (Figure 8, D). Since the system is thermally insulated during this process, the process is represented in Figure 7 by the adiabatic segment BD. During this adiabatic expansion, the temperature of the fluid decreases from t_2 to t_1, and the state of the system is now given by the point D in Figure 7.

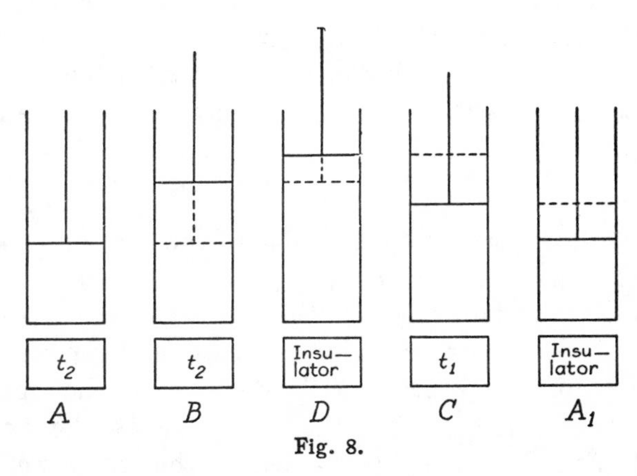

Fig. 8.

Placing the cylinder on the source t_1, we now compress the fluid very slowly along the isothermal DC (Figure 7) until its volume has decreased to V_C (Figure 8, C). Finally, we place the cylinder on the thermal insulator again and very slowly compress the fluid adiabatically along the segment CA until its temperature has increased to t_2. The system will now be at its initial state again, which is given by the point A in Figure 7 (Figure 8, A).

During the isothermal expansion represented by the segment AB, the system absorbs an amount of heat Q_2 from the source t_2. During the isothermal compression represented by the segment DC, the system absorbs an amount of heat $-Q_1$ from the source t_1; that is, it gives up an amount of heat Q_1 to the source t_1. Thus, the total amount of heat absorbed by the system during the cycle is $Q_2 - Q_1$. Let L be the amount of work done by the system during the transformation. This work is equal to the area bounded by the cycle in Figure 7. Making use of equation (16), which expresses the first law of thermodynamics for a cycle, we have:

$$L = Q_2 - Q_1 . \tag{43}$$

This equation tells us that only part of the heat that is absorbed by the system from the source at the higher temperature is transformed into work by the Carnot cycle; the rest of the heat, Q_1, instead of being transformed into work, is surrendered to the source at the lower temperature.

We define the *efficiency* of the Carnot cycle as the ratio,

$$\eta = \frac{L}{Q_2} = \frac{Q_2 - Q_1}{Q_2} = 1 - \frac{Q_1}{Q_2}, \tag{44}$$

of the work performed by the cycle to the heat absorbed at the high temperature source.

Since the Carnot cycle is reversible, it can be carried out in the reverse direction. This can be done by performing all the transformations described above in the opposite sense. When this is done, the cycle absorbs the work L instead of producing it; and it absorbs the amount of heat Q_1 at the temperature t_1 and gives up the amount of heat Q_2 at the temperature t_2.

As a first application of the Carnot cycle, we shall complete the proof of the equivalence of the Clausius and the Kelvin postulates by showing that if the Clausius postulate were not valid, Kelvin's postulate would not be valid either.

Let us assume, in contradiction to Clausius' postulate, that it were possible to transfer a certain amount of heat Q_2 from a source at the temperature t_1 to a source at a higher temperature t_2 in such a way that no other change in the state of the system occurred. With the aid of a Carnot cycle, we could then absorb this amount of heat Q_2 and produce an amount of work L. Since the source at the temperature t_2 receives and gives up the same amount of heat, it suffers no final change. Thus, the process just described would have as its only final result the transformation into work of heat extracted from a source which is at the same temperature t_1 throughout. This is contrary to the Kelvin postulate.

9. The absolute thermodynamic temperature. In the preceding section we described a reversible cyclic engine, the Carnot cycle, which performs an amount of work L during each of its cycles by absorbing a quantity of heat Q_2 from a source at the temperature t_2 and surrendering a quantity of heat Q_1 to a source at the lower temperature t_1. We shall say that such an engine works between the temperatures t_1 and t_2.

Consider now an engine working between the temperatures t_1 (lower) and t_2 (higher). Let L be the work performed by the engine during each cycle, and let Q_2 and Q_1 be the amounts of heat per cycle absorbed at the temperature t_2 and expelled at the temperature t_1, respectively. This engine need not be a Carnot cycle; the only condition we impose on it is that it be cyclic: at the end of the process it must return to its initial state.

We can easily show that if $L > 0$, that is, if the engine performs a positive amount of work, then $Q_2 > 0$ and $Q_1 > 0$.

Let us assume first that $Q_1 \leqq 0$. This would mean that the engine absorbed an amount of heat Q_1 from the source t_1 during the cycle. We could then place the two sources in thermal contact and let heat flow spontaneously by conduction from the hotter source t_2 to the colder source t_1 until the latter had received exactly the same amount of heat as it had surrendered to the engine during the cycle. Since the source t_1 would thus remain unaffected, and the engine would be back in its initial state, the only final result of this process would be the transformation into work L of heat absorbed from a single source which was initially at the same temperature t_2 throughout. Since this is in contradiction to Kelvin's postulate, we must have $Q_1 > 0$.

The proof that $Q_2 > 0$ is now very simple. Since our engine reverts to its initial state after the cycle, we have from the first law (see equation (16)):

$$L = Q_2 - Q_1 .$$

But $L > 0$ by assumption, and we have already proved that $Q_1 > 0$; hence, we must have $Q_2 > 0$.

We consider now a second engine working between the same temperatures t_1 and t_2 for which L', Q'_2, and Q'_1 are the quantities corresponding to L, Q_2, and Q_1 for the first engine. We shall prove the following fundamental theorem:

a. If the first engine is a reversible one,[2] *then,*

$$\frac{Q_2}{Q_1} \geqq \frac{Q'_2}{Q'_1} . \tag{45}$$

b. If the second engine also is reversible, then,

$$\frac{Q_2}{Q_1} = \frac{Q'_2}{Q'_1} \tag{46}$$

In part (*a*) of the theorem, we make no assumption whatever about the second engine; thus, it may or may not be reversible.

If we apply equation (16) (the special form of the first law for a cycle) to our two engines, we see that the work performed by each engine during a cycle must be equal to the difference between the heat received from the source t_2 and the heat given up at the source t_1. Thus, we must have:

$$L = Q_2 - Q_1 , \tag{47}$$

and

$$L' = Q'_2 - Q'_1 . \tag{48}$$

[2] By a "reversible" engine we mean one which operates around a reversible cycle.

The ratio Q_2/Q_2' can certainly be approximated by a rational number to as high an accuracy as we may wish. We may therefore place

$$\frac{Q_2}{Q_2'} = \frac{N'}{N}, \tag{49}$$

where N and N' are positive integers.

We now consider a process consisting of N' cycles of the second engine and N reverse cycles of the first engine. This is a permissible process, since we have assumed that the first engine is reversible. When operated in the reverse sense, the first engine absorbs an amount of work L during each reverse cycle, giving up an amount of heat Q_2 to the source t_2 and absorbing an amount of heat Q_1 from the source t_1.

The total work performed by the two engines during the complex process described above is:

$$L_{\text{total}} = N'L' - NL.$$

The total amount of heat absorbed from the source t_2 is:

$$Q_{2,\,\text{total}} = N'Q_2' - NQ_2;$$

and the total amount of heat given up to the source t_1 is:

$$Q_{1,\,\text{total}} = N'Q_1' - NQ_1.$$

From (47) and (48) we obtain immediately:

$$L_{\text{total}} = Q_{2,\,\text{total}} - Q_{1,\,\text{total}}.$$

But from (49) we deduce that:

$$Q_{2,\,\text{total}} = 0. \tag{50}$$

Hence,

$$L_{\text{total}} = -Q_{1,\,\text{total}}. \tag{51}$$

Equation (50) states that the complete process produces no exchange of heat at the high temperature t_2; and equation (51) states that the heat absorbed from the source t_1 (equal to $-Q_{1,\,\text{total}}$) is transformed into the work L_{total}.

Since the complete process is composed of several cycles of each engine, both engines will come back to their initial states at the completion of the process. From this we see that L_{total} cannot be positive; for if it were positive, the only final result of the complete process would be the transformation into work, L_{total}, of heat, $-Q_{1,\,\text{total}}$, absorbed from a source which is at the temperature t_1 throughout. But this would contradict Kelvin's postulate. Hence, we must have:

$$L_{\text{total}} \leqq 0.$$

Because of equation (51), this inequality is equivalent to

$$Q_{1,\,\text{total}} \geqq 0;$$

and remembering the expression for $Q_{1,\,\text{total}}$, we obtain:

$$N' Q_1' \geqq N Q_1 .$$

If we eliminate N' and N from this expression with the aid of equation (49), we get, since all the quantities in (49) are positive,

$$Q_2 Q_1' \geqq Q_2' Q_1 ,$$

or

$$\frac{Q_2}{Q_1} \geqq \frac{Q_2'}{Q_1'},$$

which is identical with (45).

In order to compete the proof of our fundamental theorem, we must show that if the second engine also is reversible, then the equality sign holds, as shown in equation (46).

If we take the second engine to be reversible, we have, on interchanging the two engines and applying the inequality of part (*a*) of our theorem to the new arrangement,

$$\frac{Q_2'}{Q_1'} \geqq \frac{Q_2}{Q_1}.$$

Both this inequality and (45) must hold in the present case

because both engines are reversible. But these two inequalities are compatible only if the equality sign holds.

We can restate the theorem just proved as follows:

If there are several cyclic heat engines, some of which are reversible, operating around cycles between the same temperatures t_1 *and* t_2, *all the reversible ones have the same efficiency, while the nonreversible ones have efficiencies which can never exceed the efficiency of the reversible engines.*

We consider first two reversible engines. The fact that their efficiencies are equal follows immediately from (46) and the definition (44) of efficiency.

If we have a reversible and a nonreversible engine, we obtain from the inequality (45):

$$\frac{Q_1}{Q_2} \leqq \frac{Q_1'}{Q_2'}.$$

Hence,

$$1 - \frac{Q_1}{Q_2} \geqq 1 - \frac{Q_1'}{Q_2'}.$$

Comparing this with equation (44), we see that the efficiency of the irreversible engine can never exceed that of the reversible one.

Our fundamental theorem shows us that the ratio Q_2/Q_1 has the same value for all reversible engines that operate between the same temperatures t_1 and t_2; that is, this ratio is independent of the special properties of the engine, provided it is reversible: it depends only on the temperatures t_1 and t_2. We may therefore write:

$$\frac{Q_2}{Q_1} = f(t_1, t_2), \tag{52}$$

where $f(t_1, t_2)$ is a universal function of the two temperatures t_1 and t_2.

We shall now prove that the function $f(t_1, t_2)$ has the following property:

$$f(t_1, t_2) = \frac{f(t_0, t_2)}{f(t_0, t_1)}, \tag{53}$$

where t_0, t_1, and t_2 are three arbitrary temperatures.

Let A_1 and A_2 be two reversible cyclic engines which work between the temperatures t_0 and t_1 and t_0 and t_2, respectively. If A_1 absorbs an amount of heat Q_1 at the temperature t_1 and gives up an amount of heat Q_0 at the temperature t_0 during a cycle, then from (52) we have:

$$\frac{Q_1}{Q_0} = f(t_0, t_1).$$

Similarly, if A_2 absorbs an amount of heat Q_2 at the temperature t_2 and gives up an amount of heat Q_0 at the temperature t_0 (we assume, for the sake of simplicity, that the two engines are so chosen that they give up equal amounts of heat at the temperature t_0) during each cycle, then,

$$\frac{Q_2}{Q_0} = f(t_0, t_2).$$

Dividing this equation by the preceding one, we have:

$$\frac{Q_2}{Q_1} = \frac{f(t_0, t_2)}{f(t_0, t_1)}. \tag{54}$$

Consider now a complex process consisting of a direct cycle of the engine A_2 and a reverse cycle of the engine A_1. This process is obviously a reversible cycle, since it consists of two separate reversible cycles. During the complex process no heat is exchanged at the temperature t_0, because the amount of heat Q_0 which is surrendered by the engine A_2 at the temperature t_0 is reabsorbed at that temperature by the engine A_1 operating in the reverse sense. However, at the temperature t_2 an amount of heat Q_2 is absorbed by A_2, and at the temperature t_1 an amount of heat Q_1 is expelled by the engine A_1 during the cycle. We may therefore consider A_1 and A_2, when working together in the manner described above, as forming a reversible cyclic engine which operates between the temperatures t_1 and t_2. For this engine we have, by definition of the function f:

$$\frac{Q_2}{Q_1} = f(t_1, t_2).$$

Comparing this equation with (54), we obtain (53). **Q.E.D.**

Since the temperature t_0 in the above discussion is arbitrary, we may keep it constant in all our equations; from this it follows that we may consider $f(t_0, t)$ as being a function of the temperature t only; we therefore place

$$Kf(t_0, t) = \theta(t), \tag{55}$$

where K is an arbitrary constant.

Making use of (55), we can now put (53) in the form:

$$\frac{Q_2}{Q_1} = f(t_1, t_2) = \frac{\theta(t_2)}{\theta(t_1)}. \tag{56}$$

This equation tells us that $f(t_1, t_2)$ is equal to the ratio of a function of the argument t_2 to the same function of the argument t_1.

Since we have used an empirical temperature t, it is obviously impossible to determine the analytical form of the function $\theta(t)$. Since, however, our scale of temperatures is an arbitrary one, we can conveniently introduce a new temperature scale, using θ itself as the temperature, instead of t.

It should be noticed, however, that $\theta(t)$ is not quite uniquely defined; it can be seen from (56) or (55) that $\theta(t)$ is indeterminate to the extent of an arbitrary multiplicative constant factor. We are therefore free to choose the unit of the new temperature scale θ in any way we see fit. The usual choice of this unit is made by placing the difference between the boiling temperature and the freezing temperature of water at one atmosphere of pressure equal to 100 degrees.

The temperature scale which we have just defined is called the *absolute thermodynamic scale of temperature*. It has the advantage of being independent of the special properties of any thermometric substance; furthermore, all the thermodynamic laws take on a simple form when this scale of temperature is used.

We shall now show that *the absolute thermodynamic temperature* θ *coincides with the absolute temperature* T *introduced in section* 2 *with the aid of a gas thermometer.*

We consider a Carnot cycle performed by an ideal gas (for simplicity, we take one mole of gas). Let T_1 and T_2 be the temperatures (as measured by a gas thermometer) of the two isothermals of the Carnot cycle. (See Figure 7.) We first calculate the amount of heat Q_2 absorbed at the temperature T_2 during the isothermal expansion AB. Applying the first law, equation (15), to the transformation AB, and indicating by the subscripts A and B quantities that belong to the states A and B, we have:

$$U_B - U_A + L_{AB} = Q_2,$$

where L_{AB} is the work performed during the isothermal expansion and can be calculated with the aid of equation (10):

$$L_{AB} = RT_2 \log \frac{V_B}{V_A}.$$

We now make use of the fact that the energy of an ideal gas is a function of T only (see section 5). Thus, since A and B lie on the same isothermal, we must have $U_A = U_B$, so that

$$Q_2 = L_{AB} = RT_2 \log \frac{V_B}{V_A}.$$

In a similar fashion, we can prove that the amount of heat given up at the source T_1 during the isothermal compression represented by the segment DC is:

$$Q_1 = RT_1 \log \frac{V_D}{V_C}.$$

Since the two points A and C lie on an adiabatic curve, we have, from (38):

$$T_1 V_C^{K-1} = T_2 V_A^{K-1};$$

and similarly,

$$T_1 V_D^{K-1} = T_2 V_B^{K-1}.$$

Dividing this equation by the preceding one and extracting the $(K - 1)$th root, we get:

$$\frac{V_B}{V_A} = \frac{V_D}{V_C}.$$

From this equation and the expressions for Q_2 and Q_1, we obtain:

$$\frac{Q_2}{Q_1} = \frac{T_2}{T_1}.$$

This equation shows us that the ratio Q_2/Q_1 is equal to the ratio, T_2/T_1, of the temperatures of the sources when these temperatures are expressed on the gas thermometer scale of temperature. But from (56) it follows that Q_2/Q_1 is also equal to the ratio of the temperatures of the sources when these temperatures are expressed in units of the absolute thermodynamic scale. Hence, the ratio of the two temperatures on the absolute thermodynamic scale is equal to that ratio on the gas thermometer scale; that is, the two temperature scales are proportional. Since the units of temperature for both scales have been chosen equal, we conclude that the two scales themselves are equal, that is,

$$\theta = T. \tag{57}$$

Since θ and T are equal, we need no longer use two different letters to indicate them; henceforth, we shall always use the letter T to denote the absolute thermodynamic temperature.

Using T in place of θ, we have from (56) for a reversible cycle between the temperatures T_1 and T_2:

$$\frac{Q_2}{Q_1} = \frac{T_2}{T_1}. \tag{58}$$

And the efficiency (44) of a reversible engine becomes:

$$\eta = 1 - \frac{T_1}{T_2} = \frac{T_2 - T_1}{T_2}. \tag{59}$$

10. Thermal engines. We have already proved that no engine working between two temperatures can have a higher efficiency than a reversible engine working between the same two temperatures. Thus, (59) represents the highest possible efficiency that an engine working between the temperatures T_1 and T_2 can have.

In most thermal engines the low temperature T_1 is the temperature of the environment, and is thus uncontrollable. It is therefore thermodynamically desirable to have the temperature T_2 as high as possible. Of course, we must always bear in mind the fact that the actual efficiency is generally considerably lower than the maximum efficiency (59) because all thermal engines are far from being reversible.

A Carnot cycle operated in the reverse sense can be used to extract an amount of heat Q_1 from a source at the low temperature T_1 by absorbing an amount of work L. From (43) and (58) we easily deduce that:

$$Q_1 = L \frac{T_1}{T_2 - T_1}. \tag{60}$$

On this principle we can construct a refrigerating machine using the temperature of the environment as the high temperature T_2. A Carnot cycle operated in the reverse sense could thus be used to extract the heat Q_1 from a body cooled to a temperature, T_1, lower than the temperature of the environment, T_2. It is evident from (60) that the amount of work needed to extract a given quantity of heat Q_1 from a body which is at the temperature T_1 becomes larger and larger as the temperature T_1 of the body decreases.

As in the case of an ordinary thermal engine, the efficiency of a refrigerating machine is considerably lower than the thermodynamical efficiency (60) because irreversible processes are always involved in refrigerating devices.

Problems

1. One mole of a monatomic gas performs a Carnot cycle between the temperatures 400° K and 300° K. On the upper isothermal transformation, the initial volume is 1 liter and the final volume 5 liters. To find the work performed during a cycle, and the amounts of heat exchanged with the two sources.

2. What is the maximum efficiency of a thermal engine working between an upper temperature of 400° C and a lower temperature of 18° C?

3. Find the minimum amount of work needed to extract one calorie of heat from a body at the temperature of 0° F, when the temperature of the environment is 100° F.

CHAPTER IV

The Entropy

11. Some properties of cycles. Let us consider a system S that undergoes a cyclic transformation. We suppose that during the cycle the system receives heat from or surrenders heat to a set of sources having the temperatures $T_1, T_2, \cdots, T_n$. Let the amounts of heat exchanged between the system and these sources be $Q_1, Q_2, \cdots, Q_n$, respectively; we take the Q's positive if they represent heat received by the system and negative in the other case.

We shall now prove that:

$$\sum_{i=1}^{n} \frac{Q_i}{T_i} \leqq 0, \tag{61}$$

and that the equality sign holds in (61) if the cycle is reversible.

In order to prove (61) we introduce, besides the n sources listed above, another source of heat at an arbitrary temperature T_0, and also n reversible cyclic engines (we shall take n Carnot cycles, $C_1, C_2, \cdots, C_n$) operating between the temperatures $T_1, T_2, \cdots, T_n$, respectively, and the temperature T_0. We shall choose the ith Carnot cycle, C_i, which operates between the temperatures T_i and T_0, to be of such a size that it surrenders at the temperature T_i the quantity of heat Q_i, that is, an amount equal to that absorbed by the system S at the temperature T_i.

According to (58), the amount of heat absorbed by C_i from the source T_0 is:

$$Q_{i,0} = \frac{T_0}{T_i} Q_i. \tag{62}$$

We now consider a complex cycle consisting of one cycle of the system S and one cycle of each of the Carnot cycles

$C_1, C_2, \cdots, C_n$. The net exchange of heat at each of the sources $T_1, T_2, \cdots, T_n$ during the complex cycle is zero; the source T_i surrenders an amount of heat Q_i to the system S, but it receives the same amount of heat from the cycle C_i. The source T_0, on the other hand, loses an amount of heat equal to the sum of the amounts (given by (62)) absorbed by the Carnot cycles $C_1, C_2, \cdots, C_n$. Thus, the source T_0 surrenders altogether an amount of heat equal to

$$Q_0 = \sum_{i=1}^{n} Q_{i,0} = T_0 \sum_{i=1}^{n} \frac{Q_i}{T_i}. \tag{63}$$

Hence, the net result of our complex cycle is that the system composed of S and $C_1, C_2, \cdots, C_n$ receives an amount of heat Q_0 from the source T_0. But we have already seen that in a cyclic transformation the work performed is equal to the total heat received by the system. Thus, since $S, C_1, C_2, \cdots, C_n$ return to their initial states at the end of the complex cycle, the only final result of the complex cycle is to transform into work an amount of heat received from a source at a uniform temperature T_0. If Q_0 were positive, this result would be in contradiction to Kelvin's postulate. It therefore follows that $Q_0 \leqq 0$, or, from (63),

$$\sum_{i=1}^{n} \frac{Q_i}{T_i} \leqq 0,$$

which is identical with (61).

If the cycle performed by S is reversible, we can describe it in the opposite direction, in which case all the Q_i will change sign. Applying (61) to the reverse cycle, we obtain:

$$\sum_{i=1}^{n} - \frac{Q_i}{T_i} \leqq 0,$$

or

$$\sum_{i=1}^{n} \frac{Q_i}{T_i} \geqq 0.$$

Thus, if the cycle is reversible, this inequality, as well as (61), must be satisfied. This is possible only if the equality sign holds. For a reversible cycle, therefore, we must have:

$$\sum_{i=1}^{n} \frac{Q_i}{T_i} = 0. \tag{64}$$

This completes the proof of our theorem.

In establishing (61) and (64), we assumed that the system exchanges heat with a finite number of sources $T_1, T_2, \cdots, T_n$. It is important, however, to consider the case for which the system exchanges heat with a continuous distribution of sources. In that case, the sums in (61) and (64) must be replaced by integrals extended over the entire cycle.

Denoting by $\oint$ the integral extended over a cycle and by dQ the infinitesimal amount of heat received by the system from a source at the temperature T, we have:

$$\oint \frac{dQ}{T} \leqq 0, \tag{65}$$

which is valid for all cycles, and

$$\oint \frac{dQ}{T} = 0, \tag{66}$$

which is valid only for reversible cycles.[1]

12. The entropy. The property of a reversible cycle which is expressed by (66) can also be stated in the following form. Let A and B be two equilibrium states of a system S.

[1] In order to avoid misunderstandings as to the meaning of (65) and (66), we must point out that T represents the temperature of the source which surrenders the quantity of heat dQ, and is not necessarily equal to the temperature T' of the system (or of part of the system) which receives the heat dQ. Indeed, if the cycle is irreversible (relation (65)), $T' \leqq T$ when dQ is positive, because heat cannot flow from a colder body to a hotter body; and when dQ is negative, $T' \geqq T$. If the cycle is reversible, however (equation (66)), we must always have $T' = T$, because an exchange of heat between two bodies at different temperatures is not reversible. In (66) we may therefore take T to be the temperature of the source and also the temperature of the part of the system that receives the heat dQ.

Consider a reversible transformation which takes the system from its initial state A to the final state B. In most cases many reversible transformations from A to B will be possible. For example, if the state of the system can be represented on a (V, p) diagram, any continuous curve connecting the two points A and B (representing the initial and final states of the system) corresponds to a possible reversible transformation from A to B. In Figure 9, three such transformations are shown.

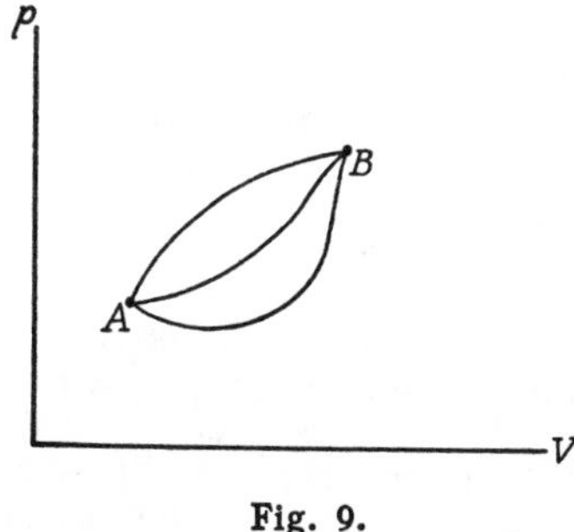

Fig. 9.

Consider now the integral:

$$\int_A^B \frac{dQ}{T}$$

extended over a reversible transformation from A to B (dQ is the amount of heat absorbed reversibly by the system at the temperature T). We shall prove that *the above integral is the same for all reversible transformations from* **A** *to* **B**; that is, *that the value of the integral for a reversible transformation depends only on the extreme states* **A** *and* **B** *of the transformation and not on the transformation itself.*

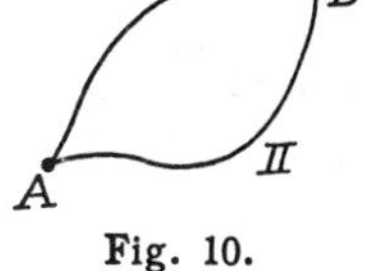

Fig. 10.

In order to prove this theorem, we must show that if I and II are two reversible transformations from A to B (in Figure 10, the states are represented as points and the transformations as lines merely as a visual aid to the proof), then,

$$\left(\int_A^B \frac{dQ}{T}\right)_{\text{I}} = \left(\int_A^B \frac{dQ}{T}\right)_{\text{II}}, \tag{67}$$

where the two integrals are taken along the paths I and II, respectively.

Consider the cyclic transformation A I B II A. This is a reversible cycle, since it is made up of two reversible transformations. We may therefore apply (66) to it, so that

$$\oint_{AIBIIA} \frac{dQ}{T} = 0.$$

This integral can be split into the sum of two integrals:

$$\left(\int_A^B \frac{dQ}{T}\right)_{\mathrm{I}} + \left(\int_B^A \frac{dQ}{T}\right)_{\mathrm{II}} = 0.$$

The second integral in this expression is equal to $-\left(\int_A^B \frac{dQ}{T}\right)_{\mathrm{II}}$, because in the transformation from B to A along II, dQ takes on the same values, except for sign, as it does in the transformation from A to B along II. Hence we obtain (67), and thus prove our theorem.

The property expressed by (67) enables us to define a new function of the state of a system. This function, which is called the *entropy* and is of utmost importance in thermodynamics, is defined in the following way:

We arbitrarily choose a certain equilibrium state O of our system and call it the *standard state*. Let A be some other equilibrium state, and consider the integral:

$$S(A) = \int_O^A \frac{dQ}{T} \tag{68}$$

taken over a reversible transformation. We have already seen that such an integral depends only on the states O and A and not on the particular reversible transformation from O to A. Since the standard state O is fixed, however, we may say that (68) is a function of the state A only. We shall call this function the entropy of the state A.[2]

[2] The necessity of restricting this definition of the entropy to equilibrium states only arises from the fact that the transformation from O to A must be reversible; that is, it must be a succession of equilibrium states. Hence it follows from continuity considerations that the initial and final states O and A must also be equilibrium states.

In many cases, however, it is possible to define the entropy even for non-equilibrium states. Let us consider, for example, a system composed of several homogeneous parts at different temperatures and pressures.

Consider now two equilibrium states A and B, and let $S(A)$ and $S(B)$, respectively, be the entropies of these states. We shall show that:

$$S(B) - S(A) = \int_A^B \frac{dQ}{T}, \tag{69}$$

where the integral is taken over a reversible transformation from state A to state B.

In order to prove this, we note that the integral on the right-hand side of (69) has the same value for all reversible transformations from A to B. We may therefore choose a particular transformation consisting of two successive reversible transformations: first a reversible transformation from A to the standard state O and then a reversible transformation from O to B. Thus, the integral in (69) can be written as the sum of two integrals:

$$\int_A^B \frac{dQ}{T} = \int_A^O \frac{dQ}{T} + \int_O^B \frac{dQ}{T}. \tag{70}$$

We have by the definition (68):

$$S(B) = \int_O^B \frac{dQ}{T},$$

since the transformation from O to B is reversible. We have further:

$$\int_A^O \frac{dQ}{T} = -\int_O^A \frac{dQ}{T} = -S(A).$$

Substituting these two values for the integrals on the right-hand side of (70), we obtain (69). **Q.E.D.**

The definition (68) of the entropy requires the arbitrary choice of a standard state O. We can easily prove that if, instead of O, we choose a different standard state O', then

---

Let each part, however, have a uniform temperature and pressure. If the different parts are in direct contact with each other, the system will evidently not be in equilibrium, since heat will flow from the hotter to the colder parts, and the differences of pressure will give rise to motion. If, however, we enclose each part in a thermally insulating rigid container, our system will be in equilibrium, and we shall be able to determine its entropy.

the new value, $S'(A)$, which we find for the entropy of the state A differs from the old one, $S(A)$, only by an additive constant.

If we take O' as the new standard state, we have, by definition,

$$S'(A) = \int_{O'}^{A} \frac{dQ}{T},$$

where the integral is extended over a reversible transformation from O' to A. By applying (69) to this integral, we find that

$$S'(A) = S(A) - S(O'),$$

or

$$S(A) - S'(A) = S(O'). \tag{71}$$

Since the new standard state O' is fixed, however, $S(O')$ is a constant (that is, it is independent of the variable state A). Thus (71) shows that the difference between the entropies of state A obtained with two different standard states, O and O', is a constant.

The entropy is thus defined except for an additive constant. This indeterminacy will not trouble us when we are dealing with entropy differences; in several problems, however, the additive constant in the entropy plays an important role. We shall see later how the third law of thermodynamics completes the definition of the entropy and also enables us to determine the entropy constant (see Chapter VIII).

Both from (68) and from (69) it follows, if we consider an infinitesimal reversible transformation during which the entropy varies by an amount dS and the system receives an amount of heat dQ at the temperature T, that

$$dS = \frac{dQ}{T}. \tag{72}$$

That is, the variation in entropy during an infinitesimal reversible transformation is obtained by dividing the amount of heat absorbed by the system by the temperature of the system.

The entropy of a system composed of several parts is very often equal to the sum of the entropies of all the parts. This is true if the energy of the system is the sum of the energies of all the parts and if the work performed by the system during a transformation is equal to the sum of the amounts of work performed by all the parts. Notice that these conditions are not quite obvious and that in some cases they may not be fulfilled. Thus, for example, in the case of a system composed of two homogeneous substances, it will be possible to express the energy as the sum of the energies of the two substances only if we can neglect the surface energy of the two substances where they are in contact. The surface energy can generally be neglected only if the two substances are not very finely subdivided; otherwise, it can play a considerable role.

Let us assume for the sake of simplicity that our system s is composed of only the two partial systems s_1 and s_2. We suppose that the energy U of s is equal to the sum of the energies U_1 and U_2 of s_1 and s_2:

$$U = U_1 + U_2;$$

and that the work L performed by s during a transformation is equal to the sum of L_1 and L_2, that is, to the sum of the work performed by s_1 and s_2, respectively:

$$L = L_1 + L_2.$$

From these assumptions and from (15) it follows that the heat Q received by the system s during a transformation can be written as the sum,

$$Q = Q_1 + Q_2,$$

of the amounts of heat received by the two parts. This enables us to split the integral (68), which defines the entropy, into the sum:

$$S(A) = \int_O^A \frac{dQ}{T} = \int_O^A \frac{dQ_1}{T} + \int_O^A \frac{dQ_2}{T},$$

of two integrals which define the entropies of the two partial systems s_1 and s_2.[3]

When the conditions for its validity are fulfilled, this additivity of entropy enables us in several cases to define the entropy of a system even though the system is not in a state of equilibrium. This is possible if we can divide the given system into a number of parts each of which alone is in a state of equilibrium. We can then define the entropy of each of these parts and, by definition, place the entropy of the total system equal to the sum of the entropies of all the parts.[4]

13. Some further properties of the entropy. Consider two states A and B of a system. We have from (69):

$$S(B) - S(A) = \int_A^B \frac{dQ}{T},$$

provided the integral is taken over a reversible transformation from A to B. If, however, the integral is taken from A to B over an irreversible transformation, the preceding equation no longer holds. We shall show in that case that we have, instead, the inequality

$$S(B) - S(A) \geqq \int_A^B \frac{dQ}{T}. \tag{73}$$

Fig. 11.

In order to show this, we take our system from A to B along an irreversible transformation, I, and back to A again along a reversible transformation R (see Figure 11). I and R together form an irreversible cycle $A\ I\ B\ R\ A$. If we apply (65) to this irreversible cycle, we obtain:

[3] It should be noticed that if the standard state O and the state A of the total system are given, the corresponding states of the two parts that compose the total system are known. These states of the two partial systems have been indicated by the same letters O and A.

[4] It can easily be proved that all the properties already shown to apply to the entropy apply also to this generalized definition.

$$0 \geqq \oint_{AIBRA} \frac{dQ}{T} = \left(\int_A^B \frac{dQ}{T}\right)_I + \left(\int_B^A \frac{dQ}{T}\right)_R.$$

Since (69) can be applied to the reversible transformation, R, from B to A, we have:

$$\left(\int_B^A \frac{dQ}{T}\right)_R = S(A) - S(B).$$

Substituting this in the preceding inequality, we obtain:

$$0 \geqq \left(\int_A^B \frac{dQ}{T}\right)_I - [S(B) - S(A)],$$

so that, for the general case of any type of transformation from A to B, we have:

$$\int_A^B \frac{dQ}{T} \leqq S(B) - S(A),$$

which is identical with (73). **Q.E.D.**

For a completely isolated system, (73) takes on a very simple form. Since for such a system $dQ = 0$, we now find that:

$$S(B) \geqq S(A); \tag{74}$$

that is, *for any transformation occurring in an isolated system, the entropy of the final state can never be less than that of the initial state.* If the transformation is reversible, the equality sign holds in (74), and the system suffers no change in entropy.

It should be clearly understood that the result (74) applies only to isolated systems. Thus, it is possible with the aid of an external system to reduce the entropy of a body. The entropy of both systems taken together, however, cannot decrease.

When an isolated system is in the state of maximum entropy consistent with its energy, it cannot undergo any further transformation because any transformation would result in a decrease of entropy. Thus, *the state of maximum entropy is the most stable state for an isolated system.* The fact that all spontaneous transformations in an isolated system proceed in such a direction as to increase the entropy can be conveniently illustrated by two simple examples.

As the first example, we consider the exchange of heat by thermal conduction between two parts, A_1 and A_2, of a system. Let T_1 and T_2 be the temperatures of these two parts, respectively, and let $T_1 < T_2$. Since heat flows by conduction from the hotter body to the colder body, the body A_2 gives up a quantity of heat Q which is absorbed by the body A_1. Thus, the entropy of A_1 changes by an amount Q/T_1, while that of A_2 changes by the amount $-Q/T_2$. The total variation in entropy of the complete system is, accordingly,

$$\frac{Q}{T_1} - \frac{Q}{T_2}.$$

Since $T_1 < T_2$, this variation is obviously positive, so that the entropy of the entire system has been increased.

As a second example, we consider the production of heat by friction. This irreversible process also results in an increase of entropy. The part of the system that is heated by friction receives a positive amount of heat and its entropy increases. Since the heat comes from work and not from another part of the system, this increase of entropy is not compensated by a decrease of entropy in another part of the system.

The fact that the entropy of an isolated system can never decrease during any transformation has a very clear interpretation from the statistical point of view. Boltzmann has proved that the entropy of a given state of a thermodynamical system is connected by a simple relationship to the probability of the state.

We have already emphasized the difference between the dynamical and thermodynamical concepts of the state of a system. To define the dynamical state, it is necessary to have the detailed knowledge of the position and motion of all the molecules that compose the system. The thermodynamical state, on the other hand, is defined by giving only a small number of parameters, such as the temperature, pressure, and so forth. It follows, therefore, that to the same thermodynamical state there corresponds a large number of dynamical states. In statistical mechanics, criteria are given for assigning to a given thermodynamical state the number π of corresponding dynamical states. (See also section 30.) This number π is usually called the *probability* of the given thermodynamical state, although, strictly speaking, it is only proportional to the probability in the usual sense. The latter can be obtained by dividing π by the total number of possible dynamical states.

We shall now assume, in accordance with statistical considerations, that in an isolated system only those spontaneous transformations occur which take the system to states of higher probability, so that the most stable state of such a system will be the state of highest probability consistent with the given total energy of the system.

We see that this assumption establishes a parallelism between the properties of the probability π and the entropy S of our system, and thus suggests the existence of a functional relationship between them. Such a relationship was actually established by Boltzmann, who proved that

$$S = k \log \pi, \tag{75}$$

where k is a constant called the *Boltzmann Constant* and is equal to the ratio,

$$\frac{R}{A}, \tag{76}$$

of the gas constant R to Avogadro's number A.

Without giving a proof of (75), we can prove, assuming the existence of a functional relationship between S and π,

$$S = f(\pi), \tag{77}$$

that the entropy is proportional to the logarithm of the probability.

Consider a system composed of two parts, and let S_1 and

S_2 be the entropies and π_1 and π_2 the probabilities of the states of these parts. We have from (77):

$$S_1 = f(\pi_1); \qquad S_2 = f(\pi_2).$$

But the entropy of the total system is the sum of the two entropies:

$$S = S_1 + S_2;$$

and the probability of the total system is the product of the two probabilities,

$$\pi = \pi_1 \pi_2.$$

From these equations and from (77) we obtain the following:

$$f(\pi_1\pi_2) = f(\pi_1) + f(\pi_2).$$

The function f must accordingly obey the functional equation:

$$f(xy) = f(x) + f(y). \tag{78}$$

This property of f enables us to determine its form. Since (78) is true for all values of x and y, we may take $y = 1 + \epsilon$, where ϵ is an infinitesimal of the first order. Then,

$$f(x + x\epsilon) = f(x) + f(1 + \epsilon).$$

Expanding both sides by Taylor's theorem and neglecting all terms of an order higher than the first, we have:

$$f(x) + x\epsilon f'(x) = f(x) + f(1) + \epsilon f'(1).$$

For $\epsilon = 0$, we find $f(1) = 0$. Hence,

$$xf'(x) = f'(1) = k,$$

where k represents a constant, or:

$$f'(x) = \frac{k}{x}.$$

Integrating, we obtain:

$$f(x) = k \log x + \text{const},$$

Remembering (77), we finally have:

$$S = k \log \pi + \text{const.}$$

We can place the constant of integration equal to zero. This is permissible because the entropy is indeterminate to the extent of an additive constant. We thus finally obtain (75).

Of course, it should be clearly understood that this constitutes no proof of the Boltzmann equation (75), since we have not demonstrated that a functional relationship between S and π exists, but have merely made it appear plausible.

14. The entropy of systems whose states can be represented on a (V, p) diagram. For these systems the state is defined by any two of the three variables, p, V, and T. If we choose T and V as the independent variables (the state variables), the heat dQ received by the system during an infinitesimal transformation as a result of which T and V change by amounts dT and dV is given by the differential expression (22)

$$dQ = \left(\frac{\partial U}{\partial T}\right)_V dT + \left[\left(\frac{\partial U}{\partial V}\right)_T + p\right] dV. \tag{79}$$

From this and (72) we obtain:

$$dS = \frac{dQ}{T} = \frac{1}{T}\left(\frac{\partial U}{\partial T}\right)_V dT + \frac{1}{T}\left[\left(\frac{\partial U}{\partial V}\right)_T + p\right] dV. \tag{80}$$

These two differential expressions for dQ and dS differ in one very important respect. We know from the general theory that there exists a function S of the state of the system. In our case, S will therefore be a function of the variables T and V, which define the state of the system:

$$S = S(T, V). \tag{81}$$

The differential expression on the right-hand side of (80) is therefore the differential of a function of the two independent variables T and V.

In general, a differential expression of two independent variables x and y, such as:

$$dz = M(x, y)dx + N(x, y)dy, \tag{82}$$

is said to be a *perfect differential* if it is the differential of a function of x and y. We may accordingly say that (80) is a perfect differential of the independent variables T and V.

It is well known that if dz is a perfect differential, then M and N must satisfy the following equation:

$$\frac{\partial M(x, y)}{\partial y} = \frac{\partial N(x, y)}{\partial x}. \tag{83}$$

When this condition is fulfilled, it is possible to integrate (82) and thus find a function which satisfies that equation. Otherwise, no such function exists, and dz cannot be considered as being the differential of some function of x and y; then, the integral of (82) along a path connecting two points on the (x, y) plane depends not only on these two points (the limits of the integral) but also on the path joining them.

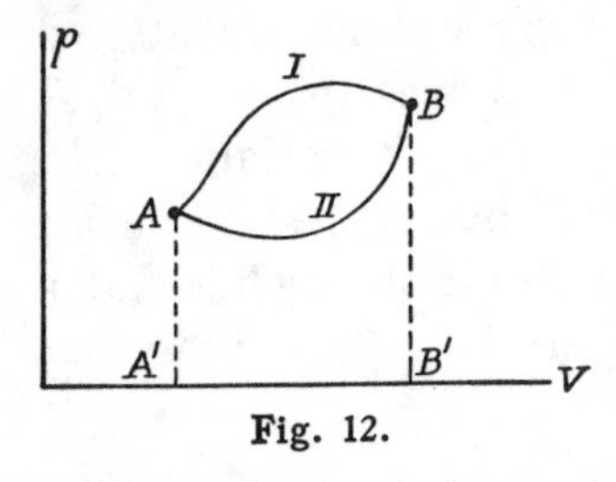

Fig. 12.

As regards the two differential expressions (79) and (80), we have already noticed that dS is a perfect differential. If we consider two states A and B on the (V, p) diagram connected by two different reversible transformations I and II (see Figure 12), and integrate dS along the two paths I and II, we get the same result in both cases, namely, $S(B) - S(A)$. If, on the other hand, we integrate dQ along these two different paths, we obtain two results, Q_1 and Q_2, which in general are not equal. This can be easily verified by applying the first law of thermodynamics, (15), to the two transformations I and II. On doing this, we find that:

$$Q_I = U(B) - U(A) + L_I$$
$$Q_{II} = U(B) - U(A) + L_{II}.$$

Taking the difference of these two expressions, we obtain:

$$Q_I - Q_{II} = L_I - L_{II}.$$

L_I and L_{II} are given by the areas $AIBB'A'A$ and $AIIBB'A'A$, respectively. Since the difference between these two areas is equal to the area $AIBIIA$, it follows that $L_I - L_{II}$ and, therefore, $Q_I - Q_{II}$ also, are, in general, different from zero. Thus, (79) is not a perfect differential, and no function Q of the state of the system can be found. It should be noticed that if a heat fluid really existed, as had been assumed before modern thermodynamics was developed, a function Q of the state of the system could be found.

Let us consider, as an example of the preceding considerations, the expressions for dQ and dS for one mole of an ideal gas. From (30) we have:

$$dQ = C_V dT + p dV,$$

or, on eliminating p with the aid of the equation of state, $pV = RT$,

$$dQ = C_V dT + \frac{RT}{V} dV. \tag{84}$$

This expression is not a perfect differential, and one can immediately verify that the condition (83) is not fulfilled.

From (84) and (72) we obtain:

$$dS = \frac{dQ}{T} = \frac{C_V}{T} dT + \frac{R}{V} dV. \tag{85}$$

Since the condition (83) is now fulfilled, this expression is a perfect differential.

On integrating (85), we obtain:

$$S = C_V \log T + R \log V + a, \tag{86}$$

where a is a constant of integration. This additive constant remains undetermined in accordance with the definition (68) of the entropy. (See, however, section 32.)

We can transform the expression (86) for the entropy of one mole of an ideal gas by introducing in place of V its value $V = RT/p$ obtained from the equation of state. Remembering (33), we obtain:

$$S = C_p \log T - R \log p + a + R \log R. \tag{87}$$

Returning to the general case of any substance whose state can be defined by the variables T and V, we obtain the expression (80) for the differential of the entropy. The condition (83), when applied to this expression, gives:

$$\frac{\partial}{\partial V}\left(\frac{1}{T}\frac{\partial U}{\partial T}\right) = \frac{\partial}{\partial T}\left[\frac{1}{T}\left(\frac{\partial U}{\partial V} + p\right)\right],$$

where we have omitted the subscripts V and T because in all these formulae we shall always use V and T as the independent variables. If we perform the partial differentiations indicated in the preceding equation and collect terms, we obtain the important result:

$$\left(\frac{\partial U}{\partial V}\right)_T = T\left(\frac{\partial p}{\partial T}\right)_V - p. \tag{88}$$

As an application of (88), we shall use it to show that the energy U of a substance which obeys the equation of state $pV = RT$ is a function of the temperature only and does not depend on the volume. We have already seen that this was experimentally verified by Joule; it is interesting, however, to obtain this result as a direct consequence of the equation of state.

Substituting the expression $p = RT/V$ in (88), we find that:

$$\begin{aligned}\left(\frac{\partial U}{\partial V}\right)_T &= T\frac{\partial}{\partial T}\left(\frac{RT}{V}\right) - \frac{RT}{V} \\ &= 0,\end{aligned}$$

which proves[5] that U does not depend on V.

If we choose T, p or p, V instead of T, V as the inde-

[5] Notice that this result is not quite independent of the Joule experiment described in section 5. Indeed, the proof of the identity between the gas thermometer temperature T and the thermodynamic temperature θ given in section 9 was based on the results of the Joule experiment.

pendent variables, we obtain two other equations which are substantially equivalent to (88). Thus, if we take T and p as the state variables, dQ is given by (23). Since $dS = dQ/T$ is a perfect differential, we easily obtain, with the aid of (83):

$$\left(\frac{\partial U}{\partial p}\right)_T = -p\left(\frac{\partial V}{\partial p}\right)_T - T\left(\frac{\partial V}{\partial T}\right)_p \tag{89}$$

Similarly, taking p and V as the independent variables, we obtain from (24) and (83):

$$T = \left[\left(\frac{\partial U}{\partial V}\right)_p + p\right]\left(\frac{\partial T}{\partial p}\right)_V - \left(\frac{\partial U}{\partial p}\right)_V\left(\frac{\partial T}{\partial V}\right)_p. \tag{90}$$

15. The Clapeyron equation. In this section we shall apply equation (88) to a saturated vapor, that is, to a system composed of a liquid and its vapor in equilibrium.

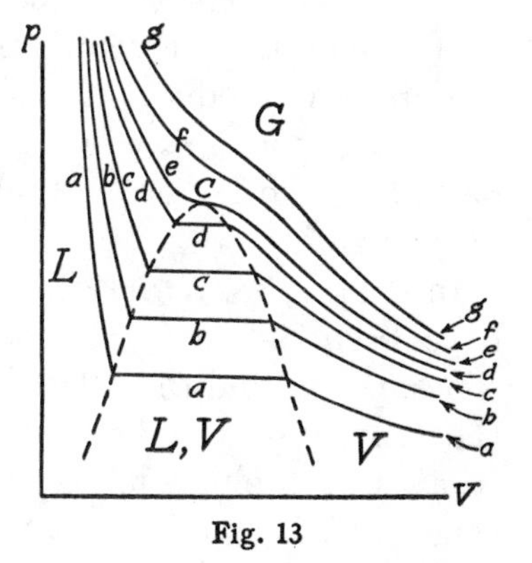

Fig. 13

We consider a liquid enclosed in a cylinder with a piston at one end. The space between the surface of the liquid and the face of the piston will be filled with saturated vapor at a pressure p which depends only on the temperature of the vapor and not on its volume.

The isothermals for this liquid-vapor system in a (V, p) representation are obtained as follows: Keeping the temperature constant, we increase the volume of the vapor by raising the piston. As a result of this, some of the liquid will evaporate in order to keep the pressure of the vapor unchanged. Thus, as long as enough liquid is left, an increase in the volume of the system leaves the pressure unchanged. Therefore, the isothermal for a mixture of a liquid and its vapor in equilibrium is a line of constant pressure, and hence parallel to the V-axis, as shown in the region within the dotted line in Figure 13.

When the volume has been increased to such an extent that all the liquid has evaporated, a further increase in volume will result, as shown in Figure 13, in a decrease in pressure just as in the case of a gas.

If we now compress our system, still keeping the temperature constant, the pressure will increase until it becomes equal to the pressure of the saturated vapor for the given temperature. At this point, a further decrease in volume does not produce an increase in the pressure; instead, some of the vapor condenses and the pressure remains unchanged (the horizontal stretch of the isothermal).

When the volume has been reduced to such an extent that the substance is completely in the liquid state, a further compression produces a very large increase in pressure, because liquids have a very low compressibility. This part of the isothermal will therefore be very steep, as shown in the figure.

In Figure 13 several isothermals of the kind just discussed have been drawn for various values of the temperature (lines a, b, c, and d). It can be seen from the figure that the length of the horizontal stretch of the isothermal (that is, the volume interval for which the liquid and vapor can coexist in equilibrium at a given temperature) decreases with increasing temperature until for the isothermal ee it reduces to an infinitesimal length (that is, to a horizontal point of inflection). This isothermal ee is called the *critical isothermal*, and its temperature T_c is called the *critical temperature.* The volume V_c and the pressure p_c corresponding to the horizontal point of inflection are called the *critical volume* and the *critical pressure*; the state corresponding to V_c, p_c, T_c is called the *critical state* (or *critical point*) of the system.

The isothermals for temperatures above the critical temperature are monotonic decreasing functions which have no discontinuities. For very large temperatures, they go over into equilateral hyperbolae, because the properties of the substance in the range of very high temperatures become more and more similar to those of an ideal gas.

The dotted line in the figure and the critical isothermal *ee* divide the (V, p) plane into four sections: the section marked L, which corresponds to the liquid state; the section marked L, V, which corresponds to the mixture of the liquid and the saturated vapor; the section V which corresponds to the nonsaturated vapor; and the section G, which corresponds to the gas.

We shall now apply (88) to the liquid-vapor system represented by region L, V of the (V, p) plane in Figure 13. In this region the pressure and the densities of the liquid and the vapor depend only on the temperature. Let v_1 and v_2 be the specific volumes (that is, the volumes per unit mass, or the inverse of the densities) of the liquid and the vapor, respectively; and let u_1 and u_2 be their specific energies (that is, the energies per unit mass). The quantities p, v_1, v_2, u_1, and u_2 are all functions of the temperature only. If m is the total mass of the substance, and m_1 and m_2 are the masses of the liquid and vapor parts, respectively, then,

$$m = m_1 + m_2 .$$

Similarly, the total volume and the total energy of the system are:

$$V = m_1 v_1(T) + m_2 v_2(T)$$

$$U = m_1 u_1(T) + m_2 u_2(T).$$

We now consider an isothermal transformation of our system which causes an amount dm of the substance to pass from the liquid state to the vapor state, and which results in a change dV of the total volume and a change dU of the total energy of the system. At the end of the transformation there will then be present $(m_1 - dm)$ grams of liquid and $(m_2 + dm)$ grams of vapor, so that the total volume will be equal to:

$$\begin{aligned} V + dV &= (m_1 - dm)v_1(T) + (m_2 + dm)v_2(T) \\ &= V + \{v_2(T) - v_1(T)\}dm, \end{aligned}$$

or

$$dV = \{v_2(T) - v_1(T)\}dm. \tag{91}$$

Similarly, the total energy will change by an amount

$$dU = \{u_2(T) - u_1(T)\}dm. \tag{92}$$

From the first law, equation (21), we have:

$$dQ = dU + pdV$$
$$= dm\{u_2 - u_1 + p(v_2 - v_1)\},$$

or

$$\frac{dQ}{dm} = u_2 - u_1 + p(v_2 - v_1) = \lambda. \tag{93}$$

Equation (93) is the expression for the amount of heat that is needed to vaporize one gram of liquid at constant temperature; it is called the latent heat of vaporization, λ. The value of λ is different for different liquids, and it also depends on the temperature. For water at the boiling temperature and standard pressure, $\lambda = 540$ cal./gm.

Since (91) and (92) refer to isothermal transformations, the ratio dU/dV gives us:

$$\left(\frac{\partial U}{\partial V}\right)_T = \frac{u_2(T) - u_1(T)}{v_2(T) - v_1(T)},$$

or, using (93):

$$\left(\frac{\partial U}{\partial V}\right)_T = \frac{\lambda}{v_2 - v_1} - p.$$

If we compare this equation with (88) and write dp/dT instead of $\left(\frac{\partial p}{\partial T}\right)_V$, which we may do because the pressure is a function of T only for our system, we find that:

$$\frac{dp}{dT} = \frac{\lambda}{T(v_2 - v_1)}. \tag{94}$$

This is called *Clapeyron's equation.*

As an example of the application of Clapeyron's equation, we shall calculate the ratio dp/dT for water vapor at the boiling temperature and at standard pressure. We have:

$$\lambda = 540 \text{ cal./gm.} = 2260 \times 10^7 \text{ ergs/gm.};$$

$$v_2 = 1677; \qquad v_1 = 1.043; \qquad T = 373.1.$$

Substituting these values in (94), we get:

$$\frac{dp}{dT} = 3.62 \times 10^4 \text{ dynes/cm.}^2 \text{ degrees} = 2.7 \text{ cm. Hg/degrees.}$$

An approximate value for dp/dT can be obtained from Clapeyron's equation by assuming that v_1 is negligible as compared to v_2, and then calculating v_2 by assuming that the vapor satisfies the equation of state of an ideal gas.

For one gram of vapor, we have, from equation (6):

$$pv_2 = \frac{R}{M} T, \tag{95}$$

where M is the molecular weight of the vapor. Equation (94) now becomes:

$$\frac{dp}{dT} = \frac{\lambda M}{RT^2} p, \tag{96}$$

or

$$\frac{d \log p}{dT} = \frac{\lambda M}{RT^2}. \tag{97}$$

For water vapor at the boiling temperature, this formula gives $dp/dT = 3.56 \times 10^4$; this is in very good agreement with the value 3.62×10^4 obtained from the exact calculation.

If the heat of vaporization λ is assumed to be constant over a wide range of temperatures, we can integrate (97) and obtain:

$$\log p = -\frac{\lambda M}{RT} + \text{constant},$$

or

$$p = \text{const. } e^{-\frac{\lambda M}{RT}} \tag{98}$$

This formula shows in a rough way how the vapor pressure depends on the temperature.

We have derived Clapeyron's equation for a liquid-vapor system, but the same formula can be applied to any change of state of a substance. As an example of this, we shall apply Clapeyron's equation to the melting of a solid. A solid subjected to a given pressure melts at a sharply defined temperature which varies with the pressure applied to the solid. Hence, for a solid-liquid system the pressure for which the solid state and the liquid state can coexist in equilibrium is a function of the temperature. We shall now use (94) to calculate the derivative of this function. The quantities λ, v_1, and v_2 in this case represent the heat of fusion and the specific volumes of the solid and the liquid, respectively.

If we take the melting of ice as an example, we have: $\lambda = 80$ cal./gm. $= 335 \times 10^7$ ergs/gm., $v_1 = 1.0907$ cm.^{3}/gm., $v_2 = 1.00013$ cm.^{3}/gm., $T = 273.1$. Substituting these values in (94), we obtain:

$$\frac{dp}{dT} = -1.35 \times 10^8 \text{ dynes/cm.}^2 \text{ degrees} = -134 \text{ atm./degrees.}$$

That is, an increase in pressure of 134 atmospheres lowers the melting point of ice by 1°.

It should be noticed, in particular, that the melting point of ice decreases with increasing pressure. In this respect water behaves differently from the way in which most substances behave; in the majority of cases, the melting point increases with increasing pressure. This anomalous behavior of water is due to the fact that ice is less dense than water, whereas in most other cases the solid is denser than the liquid.

The fact that the melting point of ice is lowered by pressure is of considerable importance in geophysics because this phenomenon is responsible for the motion of glaciers. When the mass of ice encounters a rock on the glacier bed, the high pressure of the ice against the rock lowers the melting point of the ice at that point, causing the ice to melt on one side of the rock. It refreezes again immediately after the pressure is removed. In this way the mass of ice is able to flow very slowly around obstacles.

16. The Van der Waals equation. The characteristic equation of an ideal gas represents the behavior of real gases fairly well for high temperatures and low pressures. However, when the temperature and pressure are such that the gas is near condensation, important deviations from the laws of ideal gases are observed.

Among the numerous equations of state that have been introduced to represent the behavior of real gases, that of Van der Waals is especially interesting because of its simplicity and because it satisfactorily describes the behavior of many substances over a wide range of temperatures and pressures.

Van der Waals derived his equation from considerations based on kinetic theory, taking into account to a first approximation the size of a molecule and the cohesive forces between molecules. His equation of state (written for one mole of substance) is:

$$(p + a/V^2)(V - b) = RT, \qquad (99)$$

where a and b are characteristic constants for a given substance. For $a = b = 0$, (99) reduces to the characteristic equation of an ideal gas. The term b represents the effect arising from the finite size of the molecules, and the term a/V^2 represents the effect of the molecular cohesive forces.

In Figure 14 some isothermals calculated from the Van der Waals equation of state have been drawn. If we compare them with the isothermals of Figure 13, we see that the two sets possess many similar features. In both cases there exists an isothermal having a horizontal point of inflection C. This is the critical isothermal; and the point of inflection is the critical point. The isothermals above the critical temperature show a similar behavior in both figures. However, the isothermals below the critical temperature exhibit differences. The Van der Waals isothermals are continuous curves with a minimum and a maximum, whereas the isothermals of Figure 13 have two angular points and are horizontal in the region where the Van der Waals isothermals take on their maxima and minima.

The reason for the qualitatively different behavior of the two sets of isothermals in the region marked L, V in Figure 13 is that the points on the horizontal stretch of the isothermals in Figure 13 do not correspond to homogeneous states, because along this stretch the substance splits into a liquid and a vapor part. If we compress a nonsaturated vapor isothermally until we reach the saturation pressure, and then reduce the volume still further, condensation of part of the vapor generally occurs without further increase in pressure. This corresponds to the isothermals of Figure 13. However, if we compress the vapor very gently and keep it free of dust particles, we can reach a pressure considerably higher than the saturation pressure before condensation sets in. When this situation is realized, we say that the vapor is supersaturated. The supersaturated states, however, are labile; any slight disturbance may produce condensation, causing the system to pass over into a stable state characterized by a liquid and a vapor part.

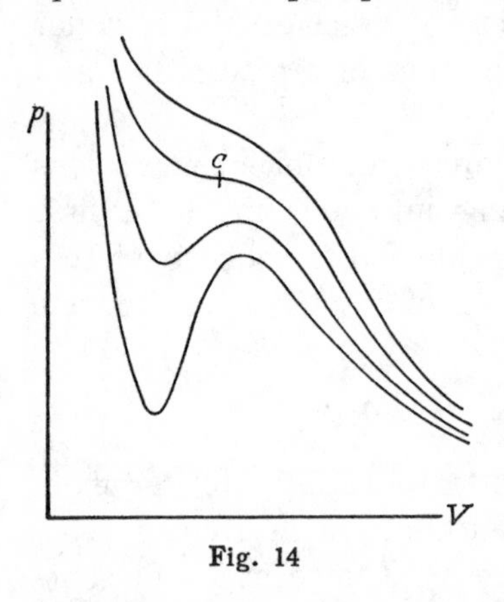

Fig. 14

The labile states are important for our discussion because they illustrate the possibility of the existence of homogeneous states in the region of the saturated vapor. We assume that these labile states are represented by the part $BCDEF$ of the Van der Waals isothermal $ABCDEFG$ (Figure 15), whereas the horizontal stretch BF of the discontinuous isothermal $ABHDIFG$ represents the stable liquid-vapor states. If it were possible to realize all the labile states on the Van der Waals isothermal, one could pass by a continuous isothermal process from the vapor represented by the part FG of the isothermal to the liquid represented by the part BA.

Given a Van der Waals isothermal, we may now wish to determine what the pressure of the saturated vapor is when its temperature is equal to that of the given isothermal; or, geometrically speaking, how high above the V-axis we must draw the horizontal stretch BF which corresponds to the liquid-vapor state. We shall prove that this distance must be such that the areas $BCDH$ and $DIFE$ are equal.

In order to prove this, we first show that the work performed by a system during a reversible isothermal cycle is always zero. From (16) we see that the work performed during a cycle is equal to the heat absorbed by the system. But for a reversible cycle, (66) holds; and since in our case the cycle is isothermal, we may remove $1/T$ from under the integral sign in (66). Equation (66) now tells us that the total heat absorbed, and, hence, the total work done during the cycle, is zero.

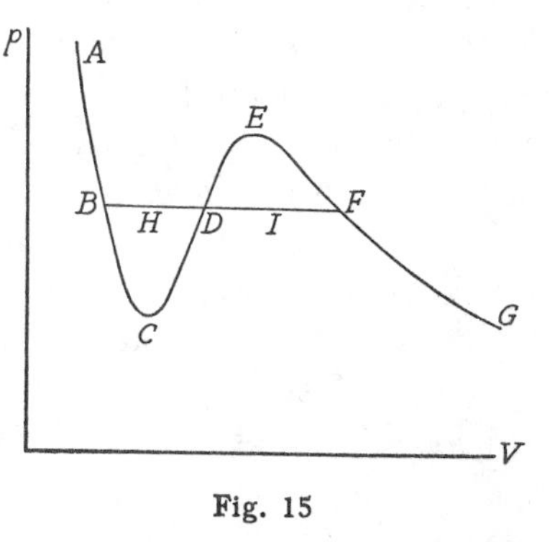

Fig. 15

We shall now consider the reversibly isothermal cycle $BCDEFIDHB$ (Figure 15). The work performed during this cycle, as measured by its area, must vanish.. But $DEFID$ is described in a clockwise direction so that its area is positive, whereas $BCDHB$, which is described in a counterclockwise direction, has a negative area. Since the total area of the cycle $BCDEFIDHB$ is zero, the absolute values of the areas of the two cycles $BCDHB$ and $DEFID$ must be equal. **Q.E.D.**

The objection might be raised against the above demonstration that since the area of the isothermal cycle *BCDHB* is obviously non-vanishing, it is not true that the work performed during a reversible isothermal cycle is always zero. The answer to this objection is that the cycle *BCDHB* is not reversible.

In order to see this, we should notice that the point D on our diagram represents two different states, depending on whether we consider it as being a point on the Van der Waals isothermal *BCDEF* or a point on the liquid-vapor isothermal *BHDIF*. Although the volume and pressure represented by D are the same in both cases, in the case of the Van der Waals isothermal, D represents a labile homogeneous state, whereas in the case of the liquid-vapor isothermal, D represents a stable nonhomogeneous state composed of a liquid and vapor part. When we perform the cycle *BCDHB*, we pass from the state D on the Van der Waals isothermal to the state D on the liquid-vapor isothermal. Since the liquid-vapor state D is more stable than the Van der Waals state D, this step is irreversible because it could not occur spontaneously in the opposite direction. Thus, the entire cycle *BCDHB* is irreversible, and therefore its area need not vanish.

The critical data T_c, V_c, and p_c of a substance can be expressed in terms of the constants a and b which appear in the Van der Waals equation of the substance.

The Van der Waals equation (99), when p and T are given, is an equation of the third degree in V. In general, therefore, there are three different roots of V for given values of T and p. The critical isothermal, $T = T_c$, however, has a horizontal point of inflection at $p = p_c$, $V = V_c$; that is, there is a third-order contact at $V = V_c$ between the critical isothermal and the horizontal line $p = p_c$. Hence, it follows that the cubic equation for V which is obtained by placing $p = p_c$ and $T = T_c$ in (99) has a triple root $V = V_c$. This cubic equation can be written in the form:

$$p_c V^3 - (p_c b + RT_c)V^2 + aV - ab = 0.$$

Since V_c is a triple root of this equation, the left-hand side must be of form $p_c(V - V_c)^3$. Hence, we find, by comparison, that:

$$V_c^3 = \frac{ab}{p_c}; \quad 3V_c^2 = \frac{a}{p_c}; \text{ and } 3V_c = \frac{p_c b + RT_c}{p_c}.$$

If we solve these three equations for V_c, p_c, and T_c, we obtain the equations:

$$V_c = 3b; \quad p_c = \frac{a}{27b^2}; \text{ and } T_c = \frac{8}{27}\frac{a}{Rb}, \tag{100}$$

which express the critical data in terms of the constants a and b.

It is worth noticing that if we take V_c, $\mathcal{P}_c$, and T_c as the units of volume, pressure, and temperature, respectively, the Van der Waals equation assumes the same form for all substances. Placing

$$\mathcal{P} = \frac{p}{p_c}; \quad \mathcal{V} = \frac{V}{V_c}; \quad \mathcal{T} = \frac{T}{T_c},$$

and making use of (100), we obtain from (99):

$$\left(\mathcal{P} + \frac{3}{\mathcal{V}^2}\right)\left(\mathcal{V} - \frac{1}{3}\right) = \frac{8}{3}\mathcal{T}. \tag{101}$$

Since this equation contains only numerical constants, it is the same for all substances. The states of various substances which are defined by the same values of $\mathcal{P}$, $\mathcal{V}$, and $\mathcal{T}$ are called *corresponding states*, and (101) is often called "Van der Waals' equation of corresponding states."

In section 14 we showed that if a substance obeys the equation of state, $pV = RT$, of an ideal gas, we can deduce thermodynamically that its energy depends on the temperature only and not on the volume. This result is true only for ideal gases. For real gases, U depends also on the volume.

From (99) we deduce that:

$$p = \frac{RT}{V - b} - \frac{a}{V^2}; \tag{102}$$

this together with (88) gives:

$$\left(\frac{\partial U}{\partial V}\right)_T = T\frac{\partial}{\partial T}\left\{\frac{RT}{V-b} - \frac{a}{V^2}\right\} - \frac{RT}{V-b} + \frac{a}{V^2}$$

$$= \frac{a}{V^2}.$$

If we integrate this equation with respect to V (keeping T constant), we obtain:

$$U = -\frac{a}{V} + f(T), \qquad (103)$$

since the constant of integration need be constant with respect to V only but may still be a function of T. The term $-\frac{a}{V}$ in (103) represents the potential energy of the cohesive forces between the molecules.

$f(T)$ cannot be further determined by means of thermodynamics alone; its determination requires some data on the specific heats. Let us assume, for example, that the molecular heat at constant volume, C_V, is constant. From (25) and (103) we obtain, then,

$$C_V = \left(\frac{\partial U}{\partial T}\right)_V = f'(T).$$

Integrating, we get:

$$f(T) = C_V T + w,$$

where w is a constant. Equation (103) now becomes:

$$U = C_V T - \frac{a}{V} + w. \qquad (104)$$

With this expression for the energy, we can easily calculate the entropy of one mole of a Van der Waals gas. From (72), (21), (102), and (104), we obtain:

$$dS = \frac{dQ}{T} = \frac{1}{T}(dU + pdV)$$

$$= \frac{1}{T}\left(C_V dT + \frac{a}{V^2}dV\right) + \frac{1}{T}\left(\frac{RT}{V-b} - \frac{a}{V^2}\right)dV$$

$$= C_V\frac{dT}{T} + R\frac{dV}{V-b},$$

or, on integrating,

$$S = C_V \log T + R \log (V - b) + \text{const.} \tag{105}$$

Notice the similarity of this formula to (86), which is the expression for the entropy of an ideal gas.

In section 6 we defined an adiabatic transformation as a reversible transformation during which the system is thermally insulated. Thus, along an adiabatic transformation $dQ = 0$, so that from (72), $dS = dQ/T = 0$, or $S =$ const. That is, if a system suffers an adiabatic transformation, its entropy remains constant. For this reason, adiabatic transformations are sometimes called *isoentropic.*

The equation of an adiabatic transformation of a Van der Waals gas is immediately obtained from (105) by taking the entropy constant. This gives:

$$C_V \log T + R \log (V - b) = \text{const.}$$

or

$$T(V - b)^{\frac{R}{C_V}} = \text{const.} \tag{106}$$

This equation for the adiabatics of a Van der Waals gas is very similar to equation (38) for the adiabatics of an ideal gas.

Problems

1. What is the entropy variation of 1,000 grams of water when raised from freezing to boiling temperature? (Assume a constant specific heat = 1 cal./gm. deg.)

2. A body obeys the equation of state:

$$pV^{1.2} = 10^9 T^{1.1}$$

A measurement of its thermal capacity inside a container having the constant volume 100 liters shows that under these conditions, the thermal capacity is constant and equal to 0.1 cal./deg. Express the energy and the entropy of the system as functions of T and V.

3. The boiling point of ethyl alcohol (C_2H_6O) is 78.3°C; the heat of vaporization is 855 joules/gm. Find dp/dT at the boiling point.

CHAPTER V

Thermodynamic Potentials

17. The free energy. In a purely mechanical system the external work L performed during a transformation is equal to minus the variation, ΔU, of its energy. That is,

$$L = -\Delta U. \tag{107}$$

For thermodynamical systems there is no such simple relationship between the work performed and the variation in energy because energy can be exchanged between the system and its environment in the form of heat. We have, instead, the first law of thermodynamics (15), which we can write in the form:

$$L = -\Delta U + Q. \tag{108}$$

Many transformations of thermodynamical systems occur while the systems are in thermal contact with the environment, so that an exchange of heat between the system and the environment can take place. In that case L may be larger or smaller than $-\Delta U$, depending on whether the system absorbs heat from or gives up heat to the environment.

We suppose now that our system s is in thermal contact with an environment which is at a constant temperature T throughout, and we consider a transformation of our system from an initial state A to a final state B. Applying the inequality (73) to this transformation, we have:

$$\int_A^B \frac{dQ}{T} \leqq S(B) - S(A).$$

Since the system receives heat only from a source whose temperature is constant, we may remove $1/T$ from under the integral sign, and we find that

$$Q = \int_A^B dQ \leqq T\{S(B) - S(A)\}. \tag{109}$$

We thus obtain an upper limit to the amount of heat which the system can receive from the environment. If the transformation from A to B is reversible, the equality sign holds in (73) and therefore in (109) also. In this case (109) gives exactly the amount of heat received by the system during the transformation.

From (108) and (109) we obtain, on putting $\Delta U = U(B) - U(A)$:

$$L \leqq U(A) - U(B) + T\{S(B) - S(A)\}. \tag{110}$$

This inequality places an upper limit on the amount of work that can be obtained during the transformation from A to B. If the transformation is reversible, the equality sign holds, and the work performed is equal to the upper limit.

Let us suppose now that the temperatures of the initial and final states, A and B, are the same and equal to the temperature T of the environment. We define a function F of the state of the system as follows:

$$F = U - TS. \tag{111}$$

In terms of this function F, which is called the *free energy* of the system, we can write (110) in the form:

$$L \leqq F(A) - F(B) = -\Delta F. \tag{112}$$

In (112), also, the equality sign holds if the transformation is reversible.

The content of equation (112) can be stated in words as follows:

If a system suffers a reversible transformation from an initial state A to a final state B both of which states have a temperature equal to that of the environment, and if the system exchanges heat with the environment only, during the transformation, the work done by the system during the transformation is equal to the decrease in the free energy of the system. If the transformation is irreversible, the decrease in the free energy of the system is only an upper limit on the work performed by the system.[1]

By comparing (112) with (107), which is true for purely mechanical systems only, we see that the free energy, in thermodynamical systems which can exchange heat with their environments, plays a role analogous to that played by the energy for mechanical systems. The main difference is that in (107) the equality sign always holds, whereas in (112) the equality sign holds only for reversible transformations.

We now consider a system that is dynamically (not thermally) insulated from its environment in the sense that any exchange of energy in the form of work between the system and its environment is impossible. The system can then perform only isochore transformations.

If the pressure at any instant of time is the same for all the parts of the system, and work can be performed by the system only as an effect of the forces exerted by this pressure on the walls, then the system is dynamically insulated when it is enclosed inside a container with invariable volume. Otherwise the dynamical insulation might require more complicated devices.

We assume that, although our system is dynamically insulated, it is in thermal contact with the environment and that its temperature is equal to the temperature T of the environment. For any transformation of our system, we have $L = 0$; we obtain thus from (112):

$$0 \leqq F(A) - F(B),$$

or

$$F(B) \leqq F(A). \tag{113}$$

[1] This result is very often stated as follows:

When a system undergoes an isothermal transformation, the work L performed by it can never exceed minus the variation, ΔF, of its free energy; it is equal to $-\Delta F$ if the transformation is reversible.

Our result is more general because it holds not only for isothermal transformations but also for transformations during which the system assumes temperatures different from T in the intermediate states, provided only that the exchange of heat occurs solely with the environment which is at the same temperature T throughout.

That is, if a system is in thermal contact with the environment at the temperature T, and if it is dynamically isolated in such a way that no external work can be performed or absorbed by the system, the free energy of the system cannot increase during a transformation.

A consequence of this fact is that, *if the free energy is a minimum, the system is in a state of stable equilibrium*; this is so because any transformation would produce an increase in the free energy, and this would be in contradiction to (113). In the case of mechanical systems, stable equilibrium exists if the potential energy is a minimum. Since the condition for stable equilibrium of a thermodynamical system enclosed in a rigid container and having the temperature of the environment is that the free energy be a minimum, the free energy is often called the "thermodynamic potential at constant volume." Notice, however, that, strictly speaking, the condition for the validity of (113) is not only that the volume of the container be constant but also that no external work be performed by the system. If the system is at a uniform pressure, however, the two conditions are equivalent.

We now consider an isothermal transformation, I, of a system at the temperature T from a state A to a state B, and also an isothermal transformation, II, between two states A' and B' at a temperature $T + dT$. A' is obtained from A by an infinitesimal transformation during which the temperature is raised by an amount dT while no external work is done. If the system is at a uniform pressure throughout, this can be realized if the volumes of A and A' are equal (isochore transformation). Similarly, during the infinitesimal transformation from B to B' no work is to be performed.

Let L and $L + dL$ be the maximum amounts of work that can be obtained from the transformations I and II, respectively. We have, then

$$L = F(A) - F(B) \tag{114}$$

$$L + dL = F(A') - F(B'),$$

or

$$\frac{dL}{dT} = \frac{dF(A)}{dT} - \frac{dF(B)}{dT}, \tag{115}$$

where we denote by $dF(A)$ and $dF(B)$, respectively, $F(A') - F(A)$ and $F(B') - F(B)$. But we have:

$$F(A) = U(A) - TS(A),$$

or, taking the differentials of both sides,

$$dF(A) = dU(A) - TdS(A) - dTS(A). \tag{116}$$

Since no work is performed in the transformation from A to A', the amount of heat received by the system during this infinitesimal transformation is, according to (15),

$$dQ_A = dU(A);$$

and, from (72),

$$dS(A) = \frac{dQ_A}{T} = \frac{dU(A)}{T}.$$

Equation (116) now gives:

$$\frac{dF(A)}{dT} = -S(A) = \frac{F(A)}{T} - \frac{U(A)}{T}.$$

Similarly, we obtain:

$$\frac{dF(B)}{dT} = -S(B) = \frac{F(B)}{T} - \frac{U(B)}{T}.$$

From (114) and (115) we thus find:

$$L - T\frac{dL}{dT} = -\Delta U, \tag{117}$$

where $\Delta U = U(B) - U(A)$ is the variation in energy resulting from the transformation from A to B. Equation (117) is called the *isochore of Van't Hoff* and has many useful applications.

At this point we shall derive a useful expression for the pressure of a system whose state can be represented on a (V, p) diagram. Let us consider an infinitesimal, isothermal, reversible transformation which changes the volume of the system by an amount dV. We can apply to this transformation equation (112) with the equality sign because the transformation is reversible. Since:

$$L = pdV, \quad \text{and} \quad \Delta F = \left(\frac{\partial F}{\partial V}\right)_T dV,$$

we have, from (112),

$$pdV = -\left(\frac{\partial F}{\partial V}\right)_T dV,$$

or

$$\left(\frac{\partial F}{\partial V}\right)_T = -p. \tag{118}$$

We conclude this section by giving the expression for the free energy of one mole of an ideal gas. This is immediately obtained from equations (111), (29), and (86):

$$F = C_V T + W - T(C_V \log T + R \log V + a). \tag{119}$$

If we use (87) instead of (86), we obtain the equivalent formula:

$$F = C_V T + W - T(C_p \log T - R \log p + a + R \log R). \tag{120}$$

18. The thermodynamic potential at constant pressure. In many thermodynamical transformations the pressure and the temperature of the system do not change but, instead, remain equal to the pressure and the temperature of the environment during the course of the transformation. Under such circumstances it is possible to define a function Φ of the state of the system which has the following property: if the function Φ is a minimum for a given set of values of the pressure and the temperature, then the system will be in equilibrium at the given pressure and temperature.

We consider an isothermal, isobaric transformation at the constant temperature T and the constant pressure p which takes our system from a state A to a state B. If $V(A)$ and $V(B)$ are the initial and final volumes occupied by the system, then the work performed during the transformation is:

$$L = p[V(B) - V(A)].$$

Since the transformation is isothermal, we may apply equation (112) to it; on doing this, we obtain:

$$pV(B) - pV(A) \leqq F(A) - F(B).$$

We now define a new function Φ of the state of the system as follows:

$$\Phi = F + pV = U - TS + pV. \tag{121}$$

In terms of Φ, the preceding inequality now becomes:

$$\Phi(B) \leqq \Phi(A). \tag{122}$$

The function Φ is called the *thermodynamic potential at constant pressure.* It follows from (122) that in an isobaric, isothermal transformation of a system, the thermodynamic potential at constant pressure can never increase.

We may therefore say that if the temperature and the pressure of a system are kept constant, *the state of the system for which the thermodynamic potential* Φ *is a minimum is a state of stable equilibrium.* The reason for this is that if Φ is a minimum, any spontaneous change in the state of the system would have the effect of increasing Φ; but this would be in contradiction to the inequality (122).

The following properties of Φ for systems whose states can be represented on a (V, p) diagram are sometimes useful.

If we choose T and p as the independent variables and differentiate (121) with respect to p, we find that:

$$\left(\frac{\partial \Phi}{\partial p}\right)_T = \left(\frac{\partial U}{\partial p}\right)_T - T\left(\frac{\partial S}{\partial p}\right)_T + p\left(\frac{\partial V}{\partial p}\right)_T + V.$$

But from the definition of the entropy and from the first law, we have for a reversible transformation:

$$dQ = TdS = dU + pdV;$$

or, in our case, for an isothermal change of pressure:

$$T\left(\frac{\partial S}{\partial p}\right)_T = \left(\frac{\partial U}{\partial p}\right)_T + p\left(\frac{\partial V}{\partial p}\right)_T.$$

Hence, we find that:

$$\left(\frac{\partial \Phi}{\partial p}\right)_T = V. \tag{123}$$

Similarly, differentiating (121) with respect to T, we can show that:

$$\left(\frac{\partial \Phi}{\partial T}\right)_p = -S. \tag{124}$$

As an example of the usefulness of the potential Φ, we shall employ it to derive Clapeyron's equation, which we have already derived in section 15 by a different method.

We consider a system composed of a liquid and its saturated vapor enclosed in a cylinder and kept at a constant temperature and pressure. If U_1, U_2, S_1, S_2, and V_1, V_2 are the energies, entropies, and volumes of the liquid and the vapor parts, respectively, and U, S, and V are the corresponding quantities for the total system, then,

$$U = U_1 + U_2$$
$$S = S_1 + S_2$$
$$V = V_1 + V_2,$$

so that, from (121),

$$\Phi = \Phi_1 + \Phi_2,$$

where Φ_1 and Φ_2 are the potentials of the liquid and vapor parts, respectively.

Let m_1 and m_2 be the masses of the liquid part and the vapor part, respectively, and let u_1, s_1, v_1, and φ_1 and u_2, s_2, v_2, and φ_2 be the specific energies, entropies, volumes, and thermodynamic potentials of the liquid and the vapor. We have, then,

$$\Phi_1 = m_1\varphi_1$$
$$\Phi_2 = m_2\varphi_2.$$

We know from the general properties of saturated vapors that all the specific quantities u_1, u_2, s_1, s_2, v_1, and v_2 and the pressure p are functions of the temperature only. Hence, φ_1 and φ_2 are functions of T only, and we may write:

$$\Phi = m_1\varphi_1(T) + m_2\varphi_2(T).$$

We start with the system in equilibrium and perform an isothermal transformation, keeping the pressure constant so that only m_1 and m_2 can vary. Let m_1 be increased by an amount dm_1 as a result of this transformation. Then, since $m_1 + m_2 = m = \text{const.}$, m_2 will decrease by an amount dm_1. The thermodynamic potential will now be given by the expression:

$$(m_1 + dm_1)\varphi_1 + (m_2 - dm_1)\varphi_2 = \Phi + dm_1(\varphi_1 - \varphi_2).$$

Since the system was initially in a state of equilibrium, Φ must have been a minimum initially. From this and from the above equation it follows that:

$$\varphi_1 = \varphi_2,$$

or

$$(u_2 - u_1) - T(s_2 - s_1) + p(v_2 - v_1) = 0.$$

Differentiating with respect to T, we find that:

$$\frac{d}{dT}(u_2 - u_1) - T\frac{d}{dT}(s_2 - s_1) - (s_2 - s_1)$$

$$+ \frac{dp}{dT}(v_2 - v_1) + p\frac{d}{dT}(v_2 - v_1) = 0.$$

But

$$T\frac{ds}{dT} = \frac{du}{dT} + p\frac{dv}{dT}.$$

Hence, the preceding equation reduces to:

$$-(s_2 - s_1) + \frac{dp}{dT}(v_2 - v_1) = 0.$$

But $(s_2 - s_1)$ is the variation in entropy when one gram of liquid is vaporized at constant temperature; hence, it is equal to λ/T, where λ is the heat of vaporization of the substance. We thus obtain the Clapeyron equation:

$$\frac{dp}{dT} = \frac{\lambda}{T(v_2 - v_1)}.$$

We shall now write down the expression for the thermodynamic potential at constant pressure for one mole of an ideal gas. From (121), (120), the equation of state, $pV = RT$, and (33), we obtain:

$$\Phi = C_pT + W - T(C_p \log T - R \log p + a + R \log R). \quad (125)$$

19. The phase rule. When a system consists of only a single homogeneous substance, it is said to consist of only one *phase.* If a heterogeneous system is composed of several parts each of which is homogeneous in itself, the system is said to consist of as many phases as there are homogeneous parts contained in the system.

As an example of a system composed of only one phase, we may consider a homogeneous liquid (not necessarily a chemically pure substance; solutions may also be considered), a homogeneous solid, or a gas.

The following are some examples of systems that consist of two phases: a system composed of water and water vapor; a saturated solution of salt in water with some of the solid salt present; a system composed of two immiscible liquids; and so forth. In the first example, the two phases are: a liquid phase composed of water, and a gaseous phase composed of the water vapor. In the second example, the two phases are: the salt-water solution, and the solid salt. In the third example, the two phases are the two liquids.

All the specific properties of a phase (that is, all the properties referred to a unit mass of the substance constituting the phase: for example, the density, the specific heat, and so forth) depend on the temperature T, the pressure p, and the chemical constitution of the phase.

In order to define the chemical constitution of a phase, we

must give the percentage of each chemically defined substance present in the phase.

Strictly speaking, one could state that if the percentage of each chemical element (counting the total amount of the element, both free and chemically bound to other elements) were known, the percentage of the different compounds that could be formed with the given elements would be determined by the given temperature T and pressure p of the phase. Indeed, it is well known from the laws of chemistry that for any given temperature, pressure, and relative concentrations of the various elements present, chemical equilibrium will always be reached within the phase. We may therefore say that a phase is a homogeneous mixture of all the possible chemical compounds which can be formed from the chemical elements present in the phase, and that the percentage of each compound present is completely determined by T, p, and the relative concentrations of all the elements in the phase.

Consider, for example, a gaseous phase consisting of definite concentrations of hydrogen and oxygen at a given temperature and pressure. The most abundant molecules formed from hydrogen and oxygen are H_2, O_2, and H_2O (for the sake of simplicity, we neglect the rarer molecules H, O, O_3, and H_2O_2). The number of water molecules which will be formed in our gaseous mixture at a given temperature and pressure is uniquely determined, and hence the constitution of the gaseous mixture also, by the concentrations of the hydrogen and the oxygen only. Strictly speaking, we may therefore say that the independent components of a phase are the chemical elements contained in the phase (each element is to be counted as an independent component whether it is present in its elementary form or in chemical combination with other elements). However, it is known from chemical considerations that under certain conditions many chemical equilibria are realized only after a period of time that is exceedingly long as compared to ordinary time intervals. Thus, if we have a gaseous mixture of H_2 and O_2

at normal temperature and pressure, chemical equilibrium is reached when a large amount of the hydrogen and the oxygen combine to form water vapor. But the reaction

$$2H_2 + O_2 = 2H_2O$$

proceeds so slowly under normal conditions that practically no combination of hydrogen and oxygen takes place in a reasonably short period of time. Of course, the reaction would take place much more rapidly if the temperature were high enough or if a suitable catalyzer were present.

We see from the preceding discussion that in all cases for which we have a chemical compound that is formed or dissociated at an extremely slow rate, we may consider the compound itself (and not its constituent elements) as a practically independent component of the phase. If, for example, we have a gaseous phase consisting of hydrogen, oxygen, and water vapor at such a low temperature that practically no water is either formed or dissociated, we shall say that our phase contains the three independent components O_2, H_2, and H_2O (and not only the two components hydrogen and oxygen); the chemical constitution of the phase is then determined by the masses of O_2, H_2, and H_2O per unit mass of the phase.

It is clear from the above considerations that the number of independent components can be either larger or smaller than the total number of chemical elements present. In the previous example we had three independent components (H_2, O_2, and H_2O) instead of only two (H and O). On the other hand, if water vapor alone is present, we can neglect its dissociation into hydrogen and oxygen and consider the phase as consisting of only one component, the water, and not of two.

Consider now a system composed of f phases and of n independent components. Let m_{ik} be the mass of the kth component present in the ith phase. Then the distribution of the components among the various phases can be conveniently described by the array:

$$\begin{array}{l} m_{11}, m_{21}, \cdots, m_{f1} \\ m_{12}, m_{22}, \cdots, m_{f2} \\ \cdots\cdots\cdots\cdots\cdots\cdots \\ m_{1n}, m_{2n}, \cdots, m_{fn}. \end{array} \tag{126}$$

At a given temperature and pressure, the condition for equilibrium of our system is that the thermodynamic potential Φ be a minimum. This condition gives rise to a set of relations among the quantities (126).

We shall assume that the surface energy of our system is negligible, so that Φ can be put equal to the sum of the thermodynamic potentials of all the phases:

$$\Phi = \Phi_1 + \Phi_2 + \cdots + \Phi_f. \tag{127}$$

The function Φ_i depends on T, p, and the masses m_{i1}, m_{i2}, $\cdots$, m_{in} of the various components in the ith phase:

$$\Phi_i = \Phi_i(t, p, m_{i1}, \cdots, m_{in}). \tag{128}$$

The form of this function depends on the special properties of the ith phase. We notice, however, that Φ_i, considered as a function of the n variables m_{i1}, m_{i2}, $\cdots$, m_{in}, is homogeneous of the first degree. Indeed, if we change m_{i1}, m_{i2}, $\cdots$, m_{in} by the same factor K, we do not alter the constitution of our phase (since it depends only on the ratios of the m's), but increase the total mass of the phase by the factor K. Thus, Φ_i becomes multiplied by the same factor K.

If our system is to be in equilibrium at a given temperature and pressure, Φ must be a minimum. This means, analytically, that if we impose on our system an infinitesimal transformation at constant temperature and pressure, the resulting variation in Φ must vanish. We consider a transformation as a result of which an amount δm (to be considered as an infinitesimal of the first order) of the kth component is transferred from the ith to the jth phase, all the other components and phases remaining unaffected. Then, m_{ik} becomes $m_{ik} - \delta m$, and m_{jk} becomes $m_{jk} + \delta m$.

In the variation of Φ, only Φ_i and Φ_j will change. Thus, we obtain as the minimum condition:

$$\delta\Phi = \delta\Phi_i + \delta\Phi_j = \frac{\partial \Phi_j}{\partial m_{jk}}\,\delta m - \frac{\partial \Phi_i}{\partial m_{ik}}\,\delta m = 0,$$

or

$$\frac{\partial \Phi_i}{\partial m_{ik}} = \frac{\partial \Phi_j}{\partial m_{jk}}. \tag{129}$$

Since a similar equation must hold for any two phases and for any one of the components, we obtain altogether the $n(f - 1)$ equations of equilibrium:

$$\begin{aligned}
&\frac{\partial \Phi_1}{\partial m_{11}} = \frac{\partial \Phi_2}{\partial m_{21}} = \cdots = \frac{\partial \Phi_f}{\partial m_{f1}} \\
&\frac{\partial \Phi_1}{\partial m_{12}} = \frac{\partial \Phi_2}{\partial m_{22}} = \cdots = \frac{\partial \Phi_f}{\partial m_{f2}} \\
&\cdots\cdots\cdots\cdots\cdots\cdots\cdots\cdots \\
&\frac{\partial \Phi_1}{\partial m_{1n}} = \frac{\partial \Phi_2}{\partial m_{2n}} = \cdots = \frac{\partial \Phi_f}{\partial m_{fn}}
\end{aligned} \tag{130}$$

We notice that these equations depend only on the chemical constitution of each phase and not on the total amount of substance present in the phase. Indeed, since (128) is a homogeneous function of the first degree in the m's, its derivative with respect to any one of the m's is homogeneous of zero degree; that is, its derivatives depend only on the ratios of $m_{i1}, m_{i2}, \cdots, m_{in}$. From the array (126), we see that there are $(n - 1)f$ such ratios (the $n - 1$ ratios of the n variables contained in a column of (126) determine the constitution of one phase). Besides these $(n - 1)f$ variables, we also have the variables T and p in (130). We thus have a total of $2 + (n - 1)f$ variables.

The difference, v, between this number and the number, $n(f - 1)$, of equations (130) is the number of the $(n - 1)f + 2$ variables which can be chosen arbitrarily, the remaining variables then being determined by the equations

(130). We therefore call v the *degree of variability* or the *number of degrees of freedom* of the system. We have:

$$v = (n - 1)f + 2 - (f - 1)n,$$

or

$$v = 2 + n - f. \tag{131}$$

This equation, which was derived by Gibbs, expresses the *phase rule.* It says that a system composed of f phases and n independent components has a degree of variability $v = 2 + n - f$. By "degree of variability" is meant the number of variables (we take as our variables T, p, and the variables that determine the constitutions of all the phases) that can be chosen arbitrarily.

To avoid misinterpretations, one should notice that only the composition and not the total amount of each phase is considered, because thermodynamic equilibrium between two phases depends only on the constitutions and not on the total amounts of the two phases present, as shown by (129). A few examples will illustrate how the phase rule is to be applied.

Example 1. A system composed of a chemically defined homogeneous fluid. We have only one phase ($f = 1$) and one component ($n = 1$). From (131) we obtain, then, $v = 2$. Thus, we can, if we wish, choose the two variables, T and p, arbitrarily; but we then have no further possibility of varying the constitution, since our substance is a chemically defined compound. (Notice that the total amount of substance, as we have already stated, is not counted as a degree of freedom.)

Example 2. A homogeneous system composed of two chemically defined gases. Here we have one phase ($f = 1$) and two independent components ($n = 2$). From (131) it follows that $v = 3$. Indeed, we may freely choose T, p, and the ratio of the two components that determines the composition of the mixture.

Example 3. Water in equilibrium with its saturated

vapor. Here we have two phases, liquid and vapor, and only one component, so that $f = 2$ and $n = 1$. Thus, we must have $v = 1$. We can choose only the temperature arbitrarily, and the pressure will then be equal to the pressure of the saturated vapor for the given temperature. Since there is only one component, we obviously have no freedom of choice in the composition of the two phases. Notice also in this example that for a given temperature we can have equilibrium between arbitrary amounts of water and water vapor provided the pressure is equal to the saturation pressure. However, the amounts of water and water vapor are not counted as degrees of freedom.

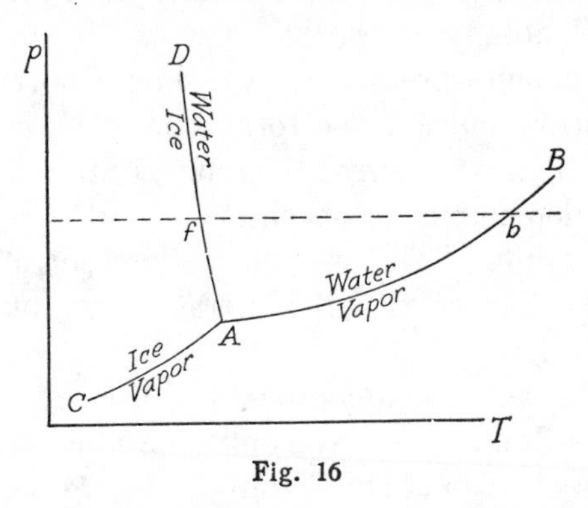

Fig. 16

Example 4. A system composed of a definite chemical compound in three different phases: solid, liquid, and vapor, as, for example, ice, water, and water vapor. We have here one component and three phases: $n = 1, f = 3$. We therefore find from (131) that $v = 0$. This means that there is no freedom of choice of the variables at all: the three phases can coexist only for a fixed value of the temperature and a fixed value of the pressure.

This fact can be illustrated with the aid of the diagram in Figure 16, in which temperatures and pressures are plotted as abscissae and ordinates, respectively.

The curve AB represents the pressure of the saturated water vapor plotted against the temperature. When the values of T and p correspond to a point on this curve, water and water vapor can coexist. If, keeping the temperature constant, we increase the pressure, equilibrium between the water and the vapor no longer exists, and all the substance condenses into the liquid phase. If, instead, we decrease the pressure, all the substance evaporates. Hence, for points above the curve AB we have water, and for points below it we have vapor, as indicated in the figure.

The curve AC is analogous to AB, but it corresponds to the pressure of the saturated vapor in contact with ice and not with liquid water. Above the curve AC ice is stable, and below it vapor is stable.

Since water and vapor can coexist along AB, and ice and vapor can coexist along AC, it is necessary that the point on the diagram corresponding to the values of T and p for which ice, water, and vapor coexist lie on both curves; that is, that this point coincide with the point of intersection A of the two curves. We see now that the three phases can coexist only for a definite value of the temperature and the pressure.

The point A is called the *triple point* because it is the intersection not only of the water-vapor curve and the ice-vapor curve but also of the ice-water curve AD. These three curves divide the T, p plane into three regions that represent the ranges of stability of vapor, ice, and water; the triple point is at the boundary of the three regions.

The triple point of water is at $T = 0.0075°C$ and $p = 0.00602$ atm. Since the pressure at the triple point is less than atmospheric pressure, the horizontal line $p = 1$ atm. (the dotted line on the diagram) intersects the three regions ice, liquid, and vapor. The intersection of the dotted line with the curve AD corresponds to a temperature equal to the freezing point f of water at atmospheric pressure (0°C). The intersection b with the curve AB corresponds to the boiling temperature of water at atmospheric pressure (100°C).

For some substances the pressure at the triple point is

higher than one atmosphere. For these substances the dotted hrizontal line corresponding to atmospheric pressure lies below the triple point and passes, therefore, directly from the solid to the vapor region without intersecting the liquid region. At atmospheric pressure these substances do not liquefy but vaporize directly from the solid phase (sublimation); they can exist in the liquid phase only at sufficiently high pressures.

20. Thermodynamics of the reversible electric cell. In all previous applications of the laws of thermodynamics, we have generally considered systems that could perform only mechanical work. But, as we have already seen in section 3, mechanical and electrical work obey the same thermodynamical laws; they are thermodynamically equivalent. The reason for this is that there are processes which can transform mechanical work completely into electrical energy, and vice versa.

As an example of a system which can perform electrical work, we shall study in this section the reversible electrolytic cell. By a "reversible electrolytic cell" we mean a cell such that a reversal of the direction of the current flowing through it causes the chemical reactions taking place in it to proceed in the opposite sense. A reversible cell can always be brought back to its initial state by reversing the flow of current through it.

Let v be the electromotive force of the cell. The electrical work performed by the cell when we permit an amount e of electricity to flow through it is:

$$L = ev. \tag{132}$$

Of course, the cell actually performs this amount of work only if we keep just a very small amount of current flowing through it, that is, if we make sure that the process occurs reversibly. Otherwise, some energy will be transformed into heat inside the cell as a result of the Joule effect.

Let $U(T)$ be the energy of our cell before any electricity

has flowed through it. We assume that $U(T)$ depends only on the temperature because we assume that the volume of our cell is practically invariable (that it is an isochore cell), and consequently neglect any possible effects which the pressure may have on the energy.

We now consider the state of the cell after a quantity e of electricity has flowed through it. The flow of electricity through the cell results in certain chemical changes within the cell, and the amount of substance which is chemically transformed is proportional to e. Thus, the energy of the cell will no longer be equal to $U(T)$ but will be changed by an amount proportional to e. Denoting by $U(T, e)$ the new energy of the cell, we have thus:

$$U(T, e) = U(T) - eu(T), \tag{133}$$

where $u(T)$ is the decrease in the energy of the cell when a unit quantity of electricity flows through it.

We now apply the Van't Hoff isochore (117) to the isothermal transformation from the initial state before any electricity has flowed through the cell (energy $= U(T)$) to the final state after the amount e has flowed through (energy given by (133)). From (133) we have for the variation in energy:

$$\Delta U = -eu(T)$$

The work performed is given by (132). Substituting in (117) and dividing both sides by e, we obtain:

$$v - T\frac{dv}{dT} = u. \tag{134}$$

This equation, which is called the *equation of Helmholtz*, establishes a relationship between the electromotive force v and the energy u. We notice that if no heat were exchanged between the cell and its environment, we should expect to find $v = u$. The extra term Tdv/dT in (134) represents the effect of the heat that is absorbed (or given out) by the cell from the environment when the electric current flows.

We can also obtain (134) directly without using the Van't Hoff isochore. Let us connect the cell to a variable condenser having a capacity C. The amount of electricity absorbed by the condenser is:

$$e = Cv(T).$$

We now consider C and T as the variables which define the state of the system composed of the cell and the condenser. If we change the capacity of the condenser by an amount dC by shifting the plates of the condenser, the system performs a certain amount of work because of the attraction between the plates. This amount of work is[2]:

$$dL = \tfrac{1}{2}\, dCv^2(T).$$

The energy of our system is the sum of the energy (133) of the cell,

$$U(T) - eu(T) = U(T) - Cv(T)u(T),$$

and the energy of the condenser, $\frac{1}{2}Cv^2(T)$. From the first law of thermodynamics (15), we find that the heat absorbed by the system in an infinitesimal transformation during which T and C change by amounts dT and dC is:

$$dQ = dU + dL = d[U(T) - Cv(T)u(T) + \tfrac{1}{2}Cv^2(T)] + \tfrac{1}{2}dC\,v^2(T)$$

$$= dT\left[\frac{dU}{dT} - Cv\frac{du}{dT} - Cu\frac{dv}{dT} + Cv\frac{dv}{dT}\right] + dC\,[v^2 - uv].$$

The differential of the entropy is, therefore,

$$dS = \frac{dQ}{T} = \frac{dT}{T}\left[\frac{dU}{dT} - Cv\frac{du}{dT} - Cu\frac{dv}{dT} + Cv\frac{dv}{dT}\right] + \frac{dC}{T}\,[v^2 - uv].$$

---

[2] This formula is obtained as follows: The energy of an isolated condenser is $\frac{1}{2}e^2/C$. If we change C, the work done is equal to minus the variation in energy, that is,

$$dL = -d\left(\frac{1}{2}\frac{e^2}{C}\right) = \frac{e^2}{2C^2}\,dC,$$

where e is kept constant because the condenser is isolated. Since $e = Cv$, we obtain the formula used in the text.

Since dS must be a perfect differential, we have:

$$\frac{\partial}{\partial C}\frac{\dfrac{dU}{dT} - Cv\dfrac{du}{dT} - Cu\dfrac{dv}{dT} + Cv\dfrac{dv}{dT}}{T} = \frac{\partial}{\partial T}\frac{v^2 - uv}{T}.$$

If we perform the differentiations indicated and remember that U, u, and v are functions of T only, we immediately obtain (134).

Problems

1. With the aid of the phase rule discuss the equilibrium of a saturated solution and the solid of the dissolved substance.

2. How many degrees of freedom has the system composed of a certain amount of water and a certain amount of air? (Neglect the rare gases and the carbon dioxide contained in air.)

3. The electromotive force of a reversible electric cell, as a function of the temperature, is:

$$0.924 + 0.0015\, t + 0.0000061\, t^2 \text{ volts},$$

t being the temperature in °C. Find the heat absorbed by the cell when one coulomb of electricity flows through it isothermally at a temperature of 18° C.

CHAPTER VI

Gaseous Reactions

21. Chemical equilibria in gases. Let us consider a gaseous system composed of a mixture of hydrogen, oxygen, and water vapor. The components of this system can interact chemically with each other according to the following chemical reaction:

$$2H_2 + O_2 \rightleftarrows 2H_2O.$$

The symbol $\rightleftarrows$ means that the reaction can proceed from left to right (formation of water) or from right to left (dissociation of water). Indeed, it is well known from the laws of chemistry that for any given temperature and pressure a state of equilibrium is reached for which the total amount of water vapor present remains unchanged, so that apparently water vapor is neither being formed nor dissociated. The actual state of affairs that exists at this equilibrium point is such that the reaction indicated above is proceeding at equal rates in both directions, so that the total amount of H_2O present remains constant. If we subtract some water vapor from the system after equilibrium has set in, the reaction from left to right will proceed with greater speed than the one from right to left until a sufficient amount of additional H_2O has been formed to establish a new state of equilibrium. If we add some water vapor, the reaction from right to left becomes preponderant for a certain length of time. Chemical equilibria in gaseous systems are regulated by the law of mass action.

We write the equation of a chemical reaction in the general form:

$$n_1A_1 + n_2A_2 + \cdots + n_rA_r \rightleftarrows m_1B_1 + m_2B_2 + \cdots + m_sB_s, \quad \textbf{(135)}$$

where $A_1, A_2, \cdots, A_r$ are the symbols for the molecules reacting on one side and $B_1, B_2, \cdots, B_s$ the symbols for those reacting on the other side. The quantities $n_1, n_2, \cdots$, and $m_1, m_2, \cdots$ are the integer coefficients of the reaction. We shall designate the concentrations of the different substances expressed in moles per unit volume by the symbols $[A_1], [A_2], \cdots$, and $[B_1], [B_2], \cdots$. We can now state the *law of mass action* as follows:

When equilibrium is reached in a chemical reaction, the expression

$$\frac{[A_1]^{n_1} [A_2]^{n_2} \cdots [A_r]^{n_r}}{[B_1]^{m_1} [B_2]^{m_2} \cdots [B_s]^{m_s}} = K(T) \tag{136}$$

is a function of the temperature only.

The quantity $K(T)$ can assume quite different values for different chemical reactions. In some cases it will be very small, and the equilibrium will be shifted toward the right-hand side; that is, when equilibrium has been reached for such cases, the concentrations of the molecules on the right-hand side are much larger than those of the molecules on the left-hand side. If, instead, $K(T)$ is large, the opposite situation exists.

It is instructive to give a very simple kinetic proof of the law of mass action. The chemical equilibrium of the reaction (135) might conveniently be called "kinetic equilibrium," because even after the equilibrium conditions have been realized, reactions among the molecules continue to take place. At equilibrium, however, the number of reactions that take place per unit time from left to right in (135) is equal to the number taking place per unit time from right to left, so that the two opposing effects compensate each other. We shall therefore calculate the number of reactions that occur per unit time from left to right and set this equal to the corresponding number of reactions proceeding in the opposite direction.

A reaction from left to right can occur as a result of a

multiple collision involving n_1 molecules A_1, n_2 molecules $A_2, \cdots, n_r$ molecules A_r. The frequency of such multiple collisions is obviously proportional to the n_1th power of $[A_1]$, to the n_2th power of $[A_2]$, $\cdots$, to the n_rth power of $[A_r]$, that is, to the product:

$$[A_1]^{n_1} [A_2]^{n_2} \cdots [A_r]^{n_r}.$$

Thus, the frequency of reactions from left to right must also be proportional to this expression. Since the temperature determines the velocities of the molecules, the proportionality factor, $K'(T)$, will be a function of the temperature. For the frequency of reactions from left to right, we obtain, then, the expression:

$$K'(T) [A_1]^{n_1} [A_2]^{n_2} \cdots [A_r]^{n_r}.$$

Similarly, for the frequency of the reactions in the opposite direction, we find:

$$K''(T) [B_1]^{m_1} [B_2]^{m_2} \cdots [B_s]^{m_s}.$$

At equilibrium these two frequencies must be equal:

$$K'(T) [A_1]^{n_1} [A_2]^{n_2} \cdots [A_r]^{n_r} = K''(T) [B_1]^{m_1} [B_2]^{m_2} \cdots [B_s]^{m_s},$$

or

$$\frac{[A_1]^{n_1} [A_2]^{n_2} \cdots [A_r]^{n_r}}{[B_1]^{m_1} [B_2]^{m_2} \cdots [B_s]^{m_s}} = \frac{K''(T)}{K'(T)}.$$

This is identical with the law of mass action (136) if we place

$$K(T) = \frac{K''(T)}{K'(T)}.$$

This simple kinetic argument gives us no information about the function $K(T)$. We shall now show that by applying thermodynamics to gaseous reactions we can not only prove the law of mass action independently of kinetic considerations, but can also determine the dependence of $K(T)$ on the temperature.

22. The Van't Hoff reaction box. The equilibria of gaseous reactions can be treated thermodynamically by assuming the existence of ideal semipermeable membranes endowed with the following two properties: (1) A membrane semipermeable to the gas A is completely impermeable to all other gases. (2) When a membrane semipermeable to the gas A separates two volumes, each containing a mixture of A and some other gas, the gas A flows through the membrane from the mixture in which its partial pressure is higher to the one in which its partial pressure is lower. Equilibrium is reached when the partial pressures of the gas A on both sides of the membrane have become equal.

Notice that a gas can flow spontaneously through a semipermeable membrane from a region of lower total pressure toward a region of higher total pressure, provided that the partial pressure of the gas that passes through the membrane is higher in the region of lower total pressure than in the region of higher total pressure. Thus, if a membrane semipermeable to hydrogen separates a box containing hydrogen at one atmosphere of pressure from a box containing oxygen at two atmospheres, hydrogen will flow through the membrane even though the total pressure on the other side is twice as large.

We should notice, finally, that in reality no ideal semipermeable membranes exist. The best approximation of such a membrane is a hot palladium foil, which behaves like a semipermeable membrane for hydrogen.

In order to study the equilibrium conditions for the chemical reaction (135), we shall first describe a process by which the reaction can be performed isothermally and reversibly. This can be done with the aid of the so-called Van't Hoff reaction box.

This box is a large container in which great quantities of the gases A_1, A_2, $\cdots$ and B_1, B_2, $\cdots$ are in chemical equilibrium at the temperature T. On one side of the box (the left side in Figure 17) is a row of r windows, the kth one of which, counting from the top down, is semipermeable to the gas A_k, while on the other side (the right-hand side of Figure 17, where we have assumed that $r = s = 2$) is a row of s windows semipermeable in the same order to the gases $B_1, B_2, \cdots, B_s$. On the outside of these windows are attached some cylinders with movable pistons, as shown in the figure.

We shall now describe a reversible, isothermal transformation of our system and calculate directly the work L performed by the system during this transformation. According to the results of section 17, however, L must be equal to the free energy of the initial state minus that of the final state of the transformation. By comparing these two expressions for L, we shall obtain the desired result.

We start with our system initially in a state for which the pistons in the cylinders, B, on the right-hand side of the

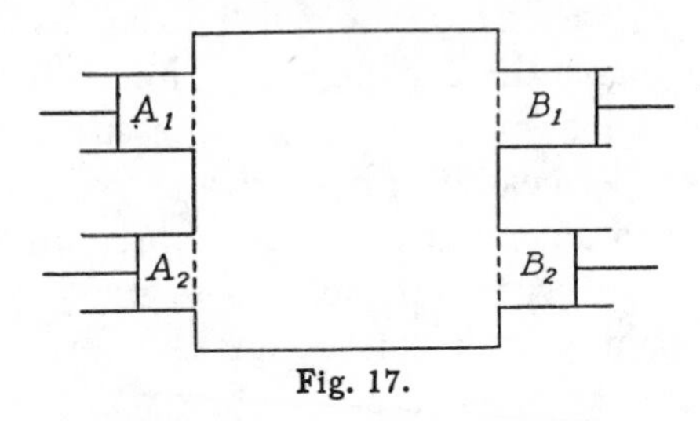

Fig. 17.

box are in contact with the windows, so that these cylinders have zero volumes, while the pistons in the r cylinders, A, on the left are in such a position that the kth cylinder contains n_k moles of the gas A_k (see Figure 18) at a concentration equal to the concentration, $[A_k]$, of this gas inside the box; the partial pressures of the gas on both sides of the semipermeable membrane are therefore equal, and a state of equilibrium exists.

The reversible transformation from the initial to the final state can be performed in the following two steps:

Step 1. Starting from the initial state (Figure 18), we shift the pistons in the cylinders on the left-hand side of the box very slowly inward until all the gases contained in these cylinders have passed through the semipermeable membranes into the large box. At the end of this process, the system will be in the intermediate state that is shown in Figure 18.

We assume that the content of the large box is so great that the relative change in concentrations resulting from this inflow of gases is negligible. The concentrations of the gases A during this process, therefore, remain practically constant and equal in order to $[A_1], [A_2] \cdots [A_r]$.

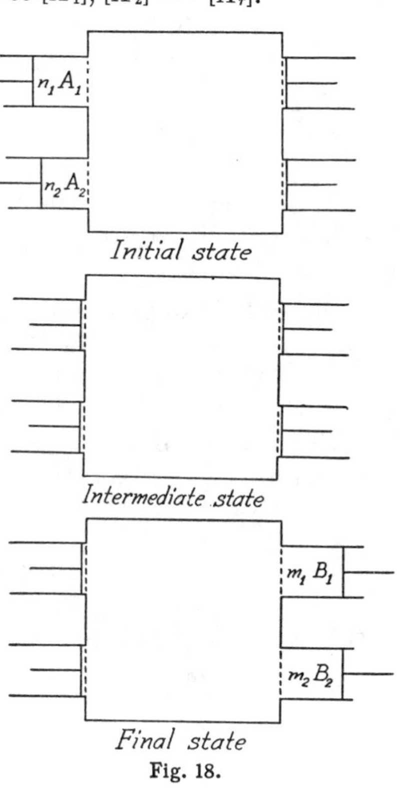

Fig. 18.

The work L performed by the system during this step is evidently negative because work must be done on the pistons against the pressures of the gases. In the first cylinder the pressure remains constant and equal to the partial pressure p_1 of the gas A_1 inside the box, while the volume of the cylinder changes from the initial volume V_1 to the final volume 0. The work is equal to the product of the constant pressure p_1 and the variation in volume, that is, $p_1(0 - V_1) = - p_1V_1$. Since the cylinder, initially, contained n_1 moles, we have, from the equation of state, $p_1V_1 = n_1RT$. The work is thus equal to $-n_1RT$. Summing the work for all the cylinders on the left, we obtain:

$$L_{\mathrm{I}} = - RT \sum_{i=1}^{r} n_i.$$

Step 2. Starting from the intermediate state, we now shift the pistons in the s cylinders on the right-hand side of the box (they are initially in contact with the windows) very slowly outward. Since the bottom of the kth cylinder, counting from the top down, is semipermeable to the gas B_k, this cylinder will absorb the gas B_k during the process and its concentration in the cylinder will be equal to that of the gas inside the large box, that is, equal to $[B_k]$. We shift the pistons outward until the cylinders, in the order from the top one down, contain $m_1, m_2, \cdots, m_s$ moles of the gases $B_1, B_2, \cdots, B_s$, respectively.

We thus reach the final state of our transformation shown on the right in Figure 18. Here the cylinders A have their pistons touching the windows so that their volumes are zero, while the pistons in the cylinders B are so placed that the kth cylinder, counting from the top down, contains m_k moles of the gas B_k at a concentration equal to the concentration, $[B_k]$, of that gas inside the box. The gases $B_1, B_2, \cdots, B_s$ in the cylinders and box are thus in equilibrium through the semipermeable bottoms of the cylinders. The work performed by the system during this second step will obviously be positive.

This work L_{II} can be calculated in the same way as in Step 1. We find:

$$L_{\text{II}} = RT \sum_{j=1}^{s} m_j.$$

The total work performed during the entire transformation is the sum of L_{I} and L_{II}, that is,

$$L = RT\left(\sum_{j=1}^{s} m_j - \sum_{i=1}^{r} n_i\right). \tag{137}$$

This work is equal to the difference between the free energy of the initial state and that of the final state. To calculate this difference, we note that the content of the large box is the same in the initial and final states. Indeed, in going from one state to the other, we first introduced into the large box n_1 moles of A_1, n_2 moles of A_2, $\cdots$, n_r moles of A_r (Step 1), and then extracted m_1 moles of B_1, m_2 moles of $B_2, \cdots, m_s$ moles of B_s. But according to the chemical equation (135), the substances introduced into the large box are equivalent to the substances withdrawn. Moreover, since the temperature and volume of the large box do not change, the chemical equilibrium of the gases in the box readjusts itself in such a way that the initial and final states of these gases are identical. The only difference between the initial and final states of the system is in the contents of the cylinders. Therefore, the difference between the free energies of the two states is equal to the difference between the free energy of the gases A contained in the cylinders A in the initial state and the free energy of the gases B contained in the cylinders B in the final state.

The free energy of the n_1 moles of A_1 in the first cylinder (initial state) can be calculated as follows: The volume occupied by one mole of the gas is evidently equal to the inverse of the concentration $[A_1]$. The free energy of one mole of A_1 is then obtained from (119) by substituting in that equation $1/[A_1]$ for the volume V of one mole. Since we have n_1 moles of A_1, the free energy of this gas is:

$$n_1\{C_{V1}T + W_1 - T(C_{V1} \log T - R \log [A_1] + a_1),$$

where C_{V1}, W_1, and a_1 are the molecular heat and the energy and entropy constants for the gas A_1. Using similar notations for $A_2, \cdots, A_r$, we find for the free energy of the gases A contained initially in the cylinders A the expression:

$$\sum_{i=1}^{r} n_i \{C_{Vi}T + W_i - T(C_{Vi} \log T - R \log [A_i] + a_i)\}$$

The free energy of the gases B in the cylinders B at the end of the process is similarly given by:

$$\sum_{j=1}^{s} m_j \{C'_{Vj}T + W'_j - T(C'_{Vj} \log T - R \log [B_j] + a'_j)\},$$

where C'_{Vj}, W'_j, and a'_j are the molecular heat and the energy and entropy constants for the gas B_j.

The difference between these two expressions must be equal to the work L given by (137). We thus have:

$$RT\left(\sum_{j=1}^{s} m_j - \sum_{i=1}^{r} n_i\right) = \sum_{i=1}^{r} n_i \{C_{Vi} T + W_i - T(C_{Vi} \log T - R \log [A_i] + a_i)\} - \sum_{j=1}^{s} m_j \{C'_{Vj} T + W'_j - T(C'_{Vj} \log T - R \log [B_j] + a'_j)\} \quad (138)$$

Dividing by RT and passing from logarithms to numbers, this equation reduces to:

$$\frac{[A_1]^{n_1}[A_2]^{n_2}\cdots[A_r]^{n_r}}{[B_1]^{m_1}[B_2]^{m_2}\cdots[B_s]^{m_s}} = e^{\frac{1}{R}\left\{\sum_{j=1}^{s} m_j(R+C'_{Vj}-a'_j) - \sum_{i=1}^{r} n_i(R+C_{Vi}-a_i)\right\}} \times T^{\frac{1}{R}\left(\sum_{i=1}^{r} C_{Vi} n_i - \sum_{j=1}^{s} C'_{Vj} m_j\right)} \times e^{-\frac{\sum_1^r n_i W_i - \sum_1^s m_j W'_j}{RT}} \quad (139)$$

The right-hand side of this equation is a function of T only. Thus, equation (139) not only proves the law of mass action (136), but it also gives the form of the function $K(T)$ explicitly.

We shall discuss the formula (139) in section 24. In the next section we shall give another proof of the same formula.

23. Another proof of the equation of gaseous equilibria. In this section we shall derive equation (139) by using the result obtained in section 17 that the states of equilibrium of a system at a given temperature and volume are those for which the free energy is a minimum.

We consider a mixture of the gases $A_1, \cdots, A_r$ and $B_1, \cdots, B_s$ at the temperature T enclosed in a container of fixed volume V and reacting chemically in accordance with equation (135). When a quantity of the gases inside the container takes part in the chemical reaction, the concentrations of the various gases present change; as a result of this, the free energy of the mixture changes also. We shall now obtain the equilibrium condition for the chemical reaction by making the free energy a minimum. To do this, we must first obtain the expression for the free energy of a mixture of gases of given concentrations.

Dalton's law (see section 2) states that the pressure of a mixture of (ideal) gases is the sum of the partial pressures of the components of the mixture (the partial pressure of a component is the pressure that this component would exert if it alone occupied the total space occupied by the mixture). This law indicates that each component is unaffected by the presence of the other components and so retains its own properties in the mixture. We shall now generalize Dalton's law by assuming that in a mixture of ideal gases the energy and the entropy also are equal to the sums of the energies and entropies (partial energies and partial entropies) which each component would have if it alone occupied the total volume occupied by the mixture at the same temperature as that of the mixture.

From the definitions (111) and (121) of the free energy and the thermodynamic potential at constant pressure, it follows now immediately that for a mixture of ideal gases these quantities are equal, respectively, to the sum of the partial free energies and the sum of the partial thermodynamic potentials at constant pressure of the components of the mixture.

With these assumptions we can now write down the expression for the free energy of our mixture of gases. The free energy of one mole of the gas A_1 is given, as in the preceding section, by the expression:

$$C_{V1}T + W_1 - T(C_{V1} \log T - R \log [A_1] + a_1).$$

Since the concentration of A_1 in the volume V is $[A_1]$, there are present altogether $V[A_1]$ moles of the gas A_1.

The partial free energy of this component of our mixture is, therefore:

$$V[A_1]\{C_{V1}T + W_1 - T(C_{V1} \log T - R \log [A_1] + a_1)\}.$$

The free energy of the total system is obtained by summing up the partial free energies of all the components in our mixture. On doing this, we obtain for the total free energy the expression:

$$F = V\sum_{i=1}^{r} [A_i]\{C_{Vi}T + W_i - T(C_{Vi} \log T - R \log [A_i] + a_i)\}$$
$$+ V\sum_{j=1}^{s} [B_j]\{C'_{Vj}T + W'_j - T(C'_{Vj} \log T - R \log [B_j] + a'_j)\} \quad (140)$$

We consider now an infinitesimal reaction of the type (135) (that is, a reaction in which an infinitesimal amount of substance is transformed). If the reaction proceeds from the left to the right of (135), infinitesimal amounts of the gases $A_1, A_2, \cdots, A_r$ disappear and infinitesimal amounts of the gases $B_1, B_2, \cdots, B_s$ are formed. The fractions of moles of the gases $A_1, A_2, \cdots, A_r$ that disappear are proportional to the coefficients $n_1, n_2, \cdots, n_r$, respectively; and the fractions of moles of the gases $B_1, B_2, \cdots, B_s$ that are produced as a result of the transformation are proportional to the numbers $m_1, m_2, \cdots, m_s$, respectively. Consequently, the concentrations $[A_1]$, $[A_2]$, $\cdots$, and $[B_1]$, $[B_2]$, $\cdots$ undergo the variations:

$$-\epsilon n_1, -\epsilon n_2, \cdots, -\epsilon n_r; \qquad \epsilon m_1, \epsilon m_2, \cdots, \epsilon m_s,$$

where ϵ is the infinitesimal constant of proportionality.

If F is to be a minimum for our state, the variation in F resulting from the infinitesimal reaction must vanish. Since this variation can be calculated as though it were a differential, we have:

$$\delta F = -\frac{\partial F}{\partial[A_1]}\epsilon n_1 - \frac{\partial F}{\partial[A_2]}\epsilon n_2 - \cdots - \frac{\partial F}{\partial[A_r]}\epsilon n_r + \frac{\partial F}{\partial[B_1]}\epsilon m_1$$
$$+ \frac{\partial F}{\partial[B_2]}\epsilon m_2 + \cdots + \frac{\partial F}{\partial[B_s]}\epsilon m_s = 0.$$

Dividing this equation by ϵV, and replacing the derivatives by their values as calculated from (140), we obtain the following equation:

$$-\sum_{i=1}^{r} n_i \{C_{Vi}T + W_i - T(C_{Vi} \log T - R \log [A_i] + a_i) + RT\}$$

$$+\sum_{j=1}^{s} m_j \{C'_{Vj}T + W'_j - T(C'_{Vj} \log T - R \log [B_j] + a'_j) + RT\} = 0.$$

It is immediately evident that this equation and equation (138) are identical. The equilibrium equation can thus be obtained at once in the same way as in the preceding section.

24. Discussion of gaseous equilibria; the principle of Le Chatelier. From (136) and (139) we can obtain the explicit form of the function $K(T)$, which appears on the right-hand side of (136). [$K(T)$ is sometimes called the *constant of the law of mass action*; of course, it is a constant only if the temperature is constant.] Comparing (136) and (139), we obtain:

$$K(T) = e^{\frac{1}{R}\left\{\sum_{j=1}^{s}(R+C'_{Vj}-a'_j)m_j - \sum_{i=1}^{r}(R+C_{Vi}-a_i)n_i\right\}}$$
$$\times T^{\frac{1}{R}\left(\sum_{i=1}^{r}C_{Vi}n_i - \sum_{j=1}^{s}C'_{Vj}m_j\right)} e^{-\frac{1}{RT}\left(\sum_{i=1}^{r}n_iW_i - \sum_{j=1}^{s}m_jW'_j\right)} \qquad (141)$$

In order to discuss the way in which $K(T)$ depends on the temperature, we first define the heat of reaction H of the chemical reaction (135). We consider a mixture of the gases A and B at constant volume and at a fixed temperature. Let these gases react according to equation (135), so that $n_1, n_2, \cdots, n_r$ moles of the gases $A_1, A_2, \cdots, A_r$, respectively, interact and give rise to $m_1, m_2, \cdots, m_s$ moles of the gases $B_1, B_2, \cdots, B_s$, respectively. The heat H developed by the system during this isothermal process is called the *heat of reaction at constant volume*. The reaction is said to be *exothermal* or *endothermal*, depending on whether heat is given out or absorbed by the system when the reaction proceeds from the left to the right in equation (135).

Since the reaction takes place at constant volume, no work is performed by the system. Therefore, the heat absorbed by the system ($= -H$) is equal, according to the first law (15), to the variation ΔU in energy of the system:

$$H = -\Delta U.$$

Remembering that the energy of one mole of A_1, for example, is equal to $C_{V1}T + W_1$, and that the numbers of moles of the gases $A_1, A_2, \cdots, A_r$ and $B_1, B_2, \cdots, B_s$ increase by the amounts $-n_1, -n_2, \cdots, -n_r$ and $m_1, m_2, \cdots, m_s$, respectively, as a result of the reaction, we find that the variation in energy associated with (135) is given by the expression:

$$\Delta U = \sum_{j=1}^{s} m_j(C'_{Vj}T + W'_j) - \sum_{i=1}^{r} n_i(C_{Vi}T + W_i).$$

The heat of reaction is thus:

$$H = \sum_{i=1}^{r} n_i(C_{Vi}T + W_i) - \sum_{j=1}^{s} m_j(C'_{Vj}T + W'_j). \tag{142}$$

Taking the logarithmic derivative of (141), we obtain:

$$\frac{d \log K(T)}{dT} = \frac{\sum_1^r C_{Vi}n_i - \sum_1^s C'_{Vj}m_j}{RT} + \frac{\sum_1^r W_i n_i - \sum_1^s W'_j m_j}{RT^2}.$$

From this equation and (142), we now find that:

$$\frac{d \log K(T)}{dT} = \frac{H}{RT^2}. \tag{143}$$

It is clear from this equation, which was derived by Helmholtz,[1] that $K(T)$ is an increasing or a decreasing function of T, depending on whether the heat of reaction is positive or negative; $K(T)$ increases with the temperature for exothermal reactions and decreases with increasing temperature for endothermal reactions.

[1] This equation can also be derived directly by applying the Van't Hoff isochore (117) to a process similar to that described in section 22.

One can easily see from (136) that an increase in $K(T)$ means a change of the equilibrium conditions in the direction of increasing concentrations of the gases A and decreasing concentrations of the gases B, that is, a shift of the equilibrium from the right to the left of equation (135). A decrease of $K(T)$, on the other hand, means that the equilibrium is shifted from the left to the right of that equation.

The effect which a change in the external conditions has on the equilibrium of a chemical reaction can best be summarized by the *Le Chatelier principle.* This principle, which enables one to determine without calculations the direction in which a change in the external conditions tends to shift the equilibrium of a thermodynamical system, states the following:

If the external conditions of a thermodynamical system are altered, the equilibrium of the system will tend to move in such a direction as to oppose the change in the external conditions.

A few examples will serve to make the meaning of this statement clear. We have already shown that if the reaction (135) is exothermal, then an increase in the temperature shifts the chemical equilibrium toward the left-hand side of equation (135). Since the reaction from left to right is exothermal, the displacement of the equilibrium toward the left results in the absorption of heat by the system and thus opposes the rise in temperature.

As a second example of the application of Le Chatelier's principle, we shall study the effect that a change in pressure (at constant temperature) has on the chemical equilibrium of the reaction (135). We notice that if the reaction (135) proceeds from left to right, then the number of moles in our gaseous system changes; if

$$n_1 + n_2 + \cdots + n_r < m_1 + m_2 + \cdots + m_s , \quad (144)$$

the number of moles increases, and if the opposite inequality holds, the number of moles decreases. If we suppose that the inequality (144) applies, then a displacement of the equilibrium toward the right will increase the pressure, and vice versa. From Le Chatelier's principle we must expect, therefore, that an increase in the pressure of our gaseous mixture will shift the equilibrium toward the left, that is, in such a direction as to oppose the increase in pressure. (In general, an increase in pressure will displace the equilibrium in such a direction as to decrease the number of moles in the system, and vice versa.) This result can be obtained directly from the law of mass action (136) as follows:

If we increase the pressure of our system while keeping the temperature constant, the concentrations of the components of our gaseous mixture increase. If the chemical equilibrium were not affected, the concentrations of all the components would be increased by the same factor, and, assuming (144) to hold, we should expect the left-hand side of (136) to decrease. But since the expression on the right-hand side of (136) remains constant, the left-hand side cannot decrease. Hence, the equilibrium must be shifted toward the left in order to keep the left-hand side of (136) constant.

We may conclude this section by stating that, in general, low pressures favor dissociation processes while high pressures favor combination processes.

Problems

1. For a chemical reaction of the type:

$$2\,A = A_2$$

the equilibrium constant $K(T)$ of the law of mass action at the temperature of 18° C is 0.00017. The total pressure of the gaseous mixture is 1 atmosphere. Find the percentage of dissociated molecules.

2. Knowing that the heat of reaction for the reaction considered in problem 1 is 50,000 cal./mole, find the degree of dissociation at 19° C and 1 atm.

CHAPTER VII

The Thermodynamics of Dilute Solutions

25. Dilute solutions. A solution is said to be *dilute* when the amount of solute is small compared to the amount of solvent. In this section we shall develop the fundamental principles of the thermodynamics of dilute solutions.

Let us consider a solution composed of N_0 moles of solvent and $N_1, N_2, \cdots, N_g$ moles of the several dissolved substances $A_1, A_2, \cdots, A_g$, respectively. If our solution is very dilute, we must have:

$$N_1 \ll N_0; N_2 \ll N_0; \cdots ; N_g \ll N_0. \tag{145}$$

Our first problem will be to find the expressions for the energy, the volume, the entropy, and so forth, of our dilute solution. A straightforward application of the thermodynamic equations will then yield all the other properties of the dilute solution.

We consider first the energy U of our solution. Let u be the energy of a fraction of the solution containing one mole of solvent. This fraction of the solution will contain N_1/N_0 moles of the solute A_1, N_2/N_0 moles of the solute $A_2, \cdots, N_g/N_0$ moles of the solute A_g. Its energy will be a function of T, p, and the quantities $N_1/N_0, N_2/N_0, \cdots, N_g/N_0$; that is,

$$u = u\left(T, p, \frac{N_1}{N_0}, \frac{N_2}{N_0}, \cdots, \frac{N_g}{N_0}\right). \tag{146}$$

Since the entire solution contains N_0 moles of solvent, its energy U is N_0 times larger than (146); that is,

$$U = N_0 u\left(T, p, \frac{N_1}{N_0}, \frac{N_2}{N_0}, \cdots, \frac{N_g}{N_0}\right). \tag{147}$$

We now make use of the fact that, since our solution is dilute, the ratios $N_1/N_0, N_2/N_0, \cdots, N_g/N_0$ are very small. We assume, therefore, that it is possible to develop the function (146) in powers of these ratios and to neglect all powers above the first. If we do this, we obtain:

$$u = u_0(T, p) + \frac{N_1}{N_0} u_1(T, p) + \frac{N_2}{N_0} u_2(T, p) + \cdots + \frac{N_g}{N_0} u_g(T, p).$$

Substituting this expression in (147), we find that:

$$\begin{aligned} U &= N_0 u_0(T, p) + N_1 u_1(T, p) + \cdots + N_g u_g(T, p) \\ &= \sum_{i=0}^{g} N_i u_i(T, p). \end{aligned} \tag{148}$$

It should be noted that although the various terms in the expression (148) for U are formally quite similar, the first term is much larger than all the others because of the inequalities (145).

By a similar process of reasoning, we can show that, to the same order of approximation, the volume can be written as:

$$\begin{aligned} V &= N_0 v_0(T, p) + N_1 v_1(T, p) + \cdots + N_g v_g(T, p) \\ &= \sum_{i=0}^{g} N_i v_i(T, p). \end{aligned} \tag{149}$$

We must now obtain the expression for the entropy of our solution. To do this, we consider an infinitesimal reversible transformation during which T and p change by the infinitesimal amounts dT and dp, while the quantities N_0, N_1, $\cdots$, N_g do not vary. The change in entropy resulting from this transformation is:

$$\begin{aligned} dS &= \frac{dQ}{T} = \frac{1}{T}(dU + pdV) \\ &= \sum_{i=0}^{g} N_i \frac{du_i + pdv_i}{T}. \end{aligned} \tag{150}$$

Since dS is a perfect differential for all values of the N's, the coefficient of each N in (150) must be a perfect differential. If we integrate these perfect differentials, we obtain a set of functions $s_0(T, p)$, $s_1(T, p)$, $\cdots$, $s_g(T, p)$ such that:

$$ds_i(T, p) = \frac{du_i + pdv_i}{T}. \tag{151}$$

If we now integrate (150), we obtain the expression for the entropy:

$$S = \sum_{i=0}^{g} N_i s_i(T, p) + C(N_0, N_1, \cdots, N_g). \tag{152}$$

The constant of integration C, which is constant only with respect to T and p, depends on the N's; we have put this in evidence in (152). We can determine the value of this constant as follows:

Since no restriction has been placed on the manner in which T and p may vary, the expression (152) for S still applies if we choose p so small and T so large that the entire solution, including all the solutes, vaporizes. Our system will then be completely gaseous, and for such a system we already know that the entropy is equal to the sum of the partial entropies of the component gases (see section 23). But the entropy of one mole of a gas at the partial pressure p_i and having the molecular heat C_{pi} is (see equation (87)):

$$C_{pi} \log T - R \log p_i + a_i + R \log R. \tag{153}$$

Hence, for our mixture of gases we have (since the partial pressure p_i of the substance A_i is equal to $pN_i/(N_0 + \cdots + N_g)$, where p is the total pressure):

$$S = \sum_{i=0}^{g} N_i \left(C_{pi} \log T - R \log p \frac{N_i}{N_0 + \cdots + N_g} + a_i + R \log R \right)$$

$$= \sum_{i=0}^{g} N_i (C_{pi} \log T - R \log p + a_i + R \log R)$$

$$- R \sum_{i=0}^{g} N_i \log \frac{N_i}{N_0 + \cdots + N_g}.$$

If we compare this with (152), which applies to our gaseous mixture also, we find that:

$$s_i = C_{pi} \log T - R \log p + a_i + R \log R,$$

and

$$C(N_0, N_1, \cdots, N_g) = -R \sum_{i=0}^{g} N_i \log \frac{N_i}{N_0 + \cdots + N_g}. \tag{154}$$

But the constant $C(N_0, N_1, \cdots, N_g)$ does not depend on T or p. Its value (154) therefore applies not only to the gaseous mixture, but also to the original solution. Hence, (152) becomes:

$$S = \sum_{i=0}^{g} N_i s_i(T, p) - R \sum_{i=0}^{g} N_i \log \frac{N_i}{N_0 + N_1 + \cdots + N_g}. \tag{155}$$

It is convenient to simplify the last term of (155) by taking the inequalities (145) into account. By neglecting terms of an order higher than the first in the small quantities $N_1, N_2, \cdots N_g$, we find that:

$$\begin{aligned} N_0 \log \frac{N_0}{N_0 + N_1 + \cdots + N_g} &= N_0 \log \frac{1}{1 + \frac{N_1}{N_0} + \cdots + \frac{N_g}{N_0}} \\ &= N_0\left(-\frac{N_1}{N_0} - \frac{N_2}{N_0} - \cdots - \frac{N_g}{N_0}\right) \\ &= -N_1 - N_2 - \cdots - N_g, \end{aligned}$$

and that:

$$N_i \log \frac{N_i}{N_0 + N_1 + \cdots + N_g} = N_i \log \frac{N_i}{N_0} \quad (\text{for } i \geqq 1).$$

Hence,

$$S = N_0 s_0(T, p) + \sum_{i=1}^{g} N_i \{s_i(T, p) + R\} - R \sum_{i=1}^{g} N_i \log \frac{N_i}{N_0}.$$

Instead of the functions s, we now introduce the new functions:

$$\begin{aligned} \sigma_0(T, p) &= s_0(T, p) \\ \sigma_1(T, p) &= s_1(T, p) + R \\ \sigma_2(T, p) &= s_2(T, p) + R \\ &\cdots\cdots\cdots\cdots \\ \sigma_g(T, p) &= s_g(T, p) + R. \end{aligned} \tag{156}$$

We have, then:

$$S = \sum_{i=0}^{g} N_i \sigma_i(T, p) - R \sum_{i=1}^{g} N_i \log \frac{N_i}{N_0}. \tag{157}$$

(Notice the difference in the limits of the two summations.)

Although the quantities u_i, v_i, and σ_i are, strictly speaking, functions of T and p, changes in these quantities resulting from variations in the pressure are very small, in general, so that u_i, v_i, σ_i, for all practical purposes, can be considered as being functions of T only.[1]

In the theory of dilute solutions we shall always make use of these approximations. We shall therefore write (148), (149), and (157) as follows:

$$U = \sum_{i=0}^{g} N_i u_i(T)$$

$$V = \sum_{i=0}^{g} N_i v_i(T)$$

$$S = \sum_{i=0}^{g} N_i \sigma_i(T) - R \sum_{i=1}^{g} N_i \log \frac{N_i}{N_0}. \tag{158}$$

With these expressions for U, V, and S, we can immediately write down the formulae for the free energy F and the thermodynamic potential Φ (see equations (111) and (121)). We have:

$$F = \sum_{i=0}^{g} N_i[u_i(T) - T\sigma_i(T)] + RT \sum_{i=1}^{g} N_i \log \frac{N_i}{N_0}$$

$$= \sum_{i=0}^{g} N_i f_i(T) + RT \sum_{i=1}^{g} N_i \log \frac{N_i}{N_0}, \tag{159}$$

where

$$f_i(T) = u_i(T) - T\sigma_i(T); \tag{160}$$

---

[1] To consider v_i as being independent of p is equivalent to neglecting the small compressibility of liquids. Similarly, u_i is very nearly independent of p; indeed, if we compress a liquid isothermally, we know from experiment that only a negligible amount of heat is developed. The work also is negligible because of the small change in volume. It follows, then, from the first law, that the variation in energy is very small. In order to show that σ_i also is practically independent of p, we observe, with the aid of (156) and (151), that:

$$\frac{\partial \sigma_i}{\partial p} = \frac{\partial s_i}{\partial p} = \frac{1}{T}\left(\frac{\partial u_i}{\partial p} + p \frac{\partial v_i}{\partial p}\right).$$

Since u_i and v_i are practically independent of p, the partial derivatives on the right-hand side are negligible. Hence, $(\partial \sigma_i/\partial p)$ is very small, and σ_i thus depends practically on T alone.

and

$$\Phi = \sum_{i=0}^{g} N_i[u_i(T) - T\sigma_i(T) + pv_i(T)] + RT \sum_{i=1}^{g} N_i \log \frac{N_i}{N_0}$$

$$= \sum_{i=0}^{g} N_i\{f_i(T) + pv_i(T)\} + RT \sum_{i=1}^{g} N_i \log \frac{N_i}{N_0}. \quad (161)$$

26. Osmotic pressure. In dealing with solutions, we shall call a semipermeable membrane a membrane that is permeable to the solvent and impermeable to the solutes. Semipermeable membranes for aqueous solutions are often found in nature. For example, the membranes of living cells are very often semipermeable. A very convenient artificial semipermeable membrane is a thin layer of copper ferrocyanide imbedded in a wall of porous material.

When a solution is separated from the pure solvent by a semipermeable membrane, a difference of pressure between the solution and the pure solvent exists at equilibrium. This can be shown by the following simple experiment.

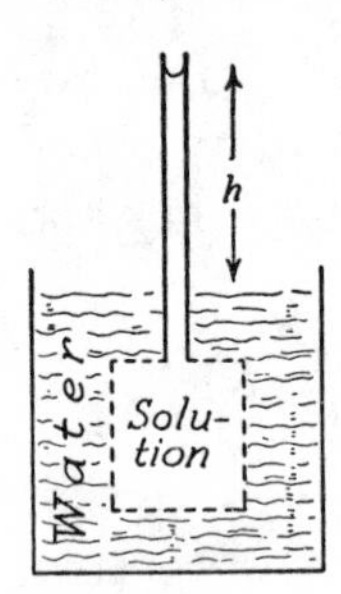

Fig. 19.

Into a container with semipermeable walls we place a solution of sugar in water. Through the top wall of the container we insert a vertical tube, as shown in Figure 19, where the semipermeable walls of the container have been indicated by dotted lines. The height of the meniscus in this tube serves to indicate the pressure of the solution inside the container. We now dip the container in a bath of pure water, and observe that the meniscus inside the tube rises above the level of the water bath. This indicates that some water has passed from the bath into the solution. Equilibrium is reached when the meniscus in the tube is at a certain height h above the level of the water bath, showing that the pressure in the solution is higher than the pressure in the pure water. The difference in pressure is called the *osmotic pressure* of the solution. If we neglect the small difference between the density of water and the density of the solution, the osmotic pressure is equal to the pressure exerted by the liquid column h, and is given by the product:

Height, h, $\times$ Density $\times$ Acceleration of Gravity.

To obtain the expression for osmotic pressure thermodynamically, we make use of the general result that the work done by a system during an isothermal reversible transformation is equal to minus the variation of the free energy. We consider the system represented in Figure 20. A cylindrical container is divided into two parts by a semipermeable membrane EF parallel to the bases AB and CD of the container. The part of the container on the left is filled with a solution composed of N_0 moles of solvent and $N_1, N_2, \cdots, N_g$ moles of several dissolved substances. The right-hand part of the container is completely filled with N_0' moles of pure solvent.

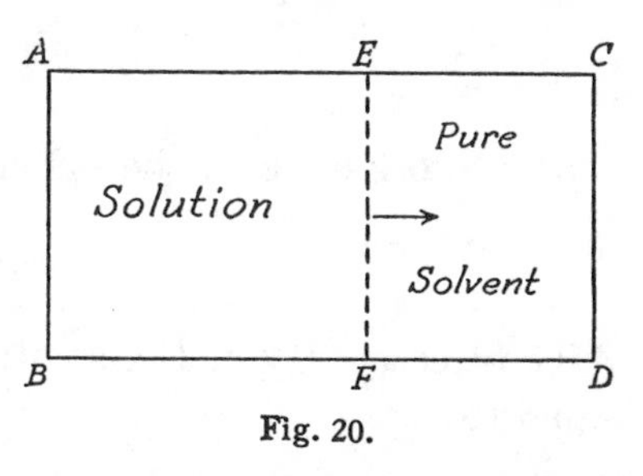

Fig. 20.

Since the membrane separating the two parts of the container is permeable to the pure solvent, there will be a flow of the pure solvent through the membrane in both directions. When these two flows become equal, the system will be in equilibrium, and there will then be a difference of pressure between the left-hand part of the container and the right-hand part. This difference of pressure P is equal to the osmotic pressure.

We assume now that the semipermeable membrane is movable, and we consider an infinitesimal transformation of our system during which the membrane is shifted an infinitesimal distance toward the right, so that the volume on the left increases by an amount dV and the volume on the right decreases by the same amount. Since the pressure exerted on the left face of the membrane by the solution is larger by an amount P than the pressure exerted on the right face of the membrane by the pure solvent, the work done by the system is PdV.

During the motion of the membrane, a certain amount (dN_0 moles) of the solvent flows from the right-hand side of the container into the solution on the left-hand side, thus diluting the solution. The volumes V and V' of the solution and the pure solvent, respectively, prior to the transformation are, according to the second of equations (158):

$$V = N_0 v_0 + N_1 v_1 + \cdots + N_g v_g$$

$$V' = N_0' v_0. \tag{162}$$

If N_0 increases by an amount dN_0, we have from the first equation[2]:

$$dV = v_0 dN_0;$$

and the work done by the system is, therefore,

$$P v_0 dN_0. \tag{163}$$

The free energy of the solution is given by (159), and is equal to:

$$N_0 f_0 + N_1 f_1 + \cdots + N_g f_g + RT\left(N_1 \log \frac{N_1}{N_0} + \cdots + N_g \log \frac{N_g}{N_0}\right).$$

The free energy of the pure solvent is obtained from this formula by replacing N_0 by N_0' and putting $N_1 = N_2 = \cdots = N_g = 0$. This gives:

$$N_0' f_0.$$

The total free energy of our system is equal to the sum of these two:

$$F = (N_0 + N_0') f_0 + N_1 f_1 + \cdots + N_g f_g + RT \sum_{i=1}^{g} N_i \log \frac{N_i}{N_0}.$$

[2] Since N_0' decreases by an amount dN_0, we have $dV' = -v_0 dN_0$, so that the total volume remains unchanged.

Since N_0 and N'_0 change by amounts dN_0 and $-dN_0$, respectively, as a result of the transformation, the variation in F is given by

$$\begin{aligned} dF &= \frac{\partial F}{\partial N_0} dN_0 - \frac{\partial F}{\partial N'_0} dN_0 \\ &= \left\{ f_0 - \frac{RT}{N_0} \sum_{i=1}^{g} N_i \right\} dN_0 - f_0 dN_0 \\ &= -\frac{RT}{N_0} dN_0 \sum_{i=1}^{g} N_i . \end{aligned}$$

The negative of this quantity must be equal to the work (163) because the transformation is reversible. Thus:

$$Pv_0 dN_0 = \frac{RT}{N_0} dN_0 \sum_{i=1}^{g} N_i ,$$

or

$$Pv_0 N_0 = RT \sum_{i=1}^{g} N_i . \tag{164}$$

$N_0 v_0$, which is the volume occupied by N_0 moles of pure solvent, differs very little from the volume V of the dilute solution (see (145) and the first of equations (162)). Neglecting this small difference[3] and replacing $N_0 v_0$ by V in (164), we obtain:

$$PV = RT \sum_{i=1}^{g} N_i , \tag{165}$$

or

$$P = \frac{RT}{V} (N_1 + N_2 + \cdots + N_g). \tag{166}$$

The above expression for the osmotic pressure of a solution bears a very close resemblance to the equation of state of a gas. Equation (166) can be stated as follows:

[3] It is immediately seen that this approximation consists in disregarding terms containing the squares of the concentrations of the solutes, and is therefore consistent with all the approximations already made in the theory of dilute solutions.

The osmotic pressure of a dilute solution is equal to the pressure exerted by an ideal gas at the same temperature and occupying the same volume as the solution and containing a number of moles equal to the number of moles of the solutes dissolved in the solution.

This simple thermodynamical result can be easily interpreted from the point of view of the kinetic theory. We consider a container divided into two parts by a semipermeable membrane with pure solvent in each part. Since the solvent can pass freely through the semipermeable membrane, the pressure on both sides of the membrane will be the same. Now let us dissolve some substances in one part and not in the other. Then the pressure on the side of the membrane facing the solution will be increased by the impacts against it of the molecules of the dissolved substances, which cannot pass through the membrane and which move about with a velocity that depends on T. The larger the number of molecules dissolved and the higher the temperature, the larger will be the number of impacts per unit time and, hence, the greater the osmotic pressure.

It can be shown from kinetic theory that the velocities of the molecules of the dissolved substances are not affected by the molecules' being in solution, but are equal to the velocities that they would have if they were in a gaseous state. Therefore, both the number and the intensity of the impacts of the molecules of the dissolved substances against the membrane are equal to the number and intensity of the impacts that one expects for a gas. The pressures exerted in both cases are therefore equal.

In order to calculate the osmotic pressure with the aid of (166), it is necessary to know the total number of moles of the dissolved substances in the solution. If no chemical change takes place in the solutes as a result of their being in solution, this number can be calculated immediately from the knowledge of the molecular weights of the solutes and the percentage by weight of these substances present in the solution. For example, a normal solution, that is, a solution containing 1 mole of solute per liter of water, has, at 15°C, an osmotic pressure:

$$P_{\text{normal}} = \frac{R \times 288.1}{1000} = 2.4 \times 10^7 \frac{\text{dynes}}{\text{cm.}^2} = 23.7 \text{ atm.}$$

In many cases, however, a chemical transformation takes place when a substance is dissolved, so that the number of moles of the substance in the solution need not be the same as the number of moles before the substance is dissolved. The most important example of this is that of an electrolyte dissolved in water. When, for example, NaCl is dissolved in water, almost all the NaCl molecules dissociate into Na^+ and Cl^- ions. The number of molecules in the solution is thus about twice the number one would expect to find if no dissociation occurred. Some electrolytes, of course, dissociate into more than two ions. For strong electrolytes, the dissociation is practically complete even when the solution is not very dilute. For the case of weak electrolytes, on the other hand, chemical equilibrium sets in between the dissociation of the electrolyte into ions and the recombination of these ions. The dissociation in this case, therefore, is generally incomplete.

27. Chemical equilibria in solutions. We have already seen that the law of mass action (136) applies to chemical reactions taking place in gaseous systems. We shall now derive a corresponding law for chemical reactions occurring in solutions.

Let A_0 represent a molecule of the solvent and $A_1, \cdots, A_r$ and $B_1, \cdots, B_s$ represent the molecules of the solutes. We assume that a chemical reaction defined by the equation:

$$n_0 A_0 + n_1 A_1 + \cdots + n_r A_r \rightleftarrows m_1 B_1 + \cdots + m_s B_s \quad (167)$$

can take place among these substances. If $n_0 \neq 0$, the solvent also takes part in the reaction; whereas if $n_0 = 0$, only the solutes react among themselves.

Just as in section 23, we shall require that when chemical

equilibrium is reached, the free energy shall be a minimum.[4] The free energy of the solution is given, according to (159), by:

$$F = f_0 N_0 + \sum_{i=1}^{r} f_i N_i + \sum_{j=1}^{s} f'_j N'_j + RT\left\{\sum_{i=1}^{r} N_i \log \frac{N_i}{N_0} + \sum_{j=1}^{s} N'_j \log \frac{N'_j}{N_0}\right\}, \quad (168)$$

where f_i and f'_j are the functions of T for the dissolved substances A_i and B_j which correspond to the functions $f_1, \cdots, f_g$ appearing in equation (159), and N_0, N_i, and N'_j are the numbers of moles of the solvent and the dissolved substances A_i and B_j, respectively.

Just as in section 23, we now consider an infinitesimal isothermal reaction of the type (167) as a result of which $N_0, N_1, \cdots, N_r$ and $N'_1, \cdots N'_s$ change by the amounts:

$$-\epsilon n_0, \ -\epsilon n_1, \ \cdots, \ -\epsilon n_r; \ \epsilon m_1, \ \cdots, \ \epsilon m_s,$$

respectively, where ϵ is an infinitesimal constant of proportionality. Since F is a minimum at equilibrium, its variation must vanish when the system is in a state of equilibrium. We thus have:

$$\delta F = -\epsilon n_0 \frac{\partial F}{\partial N_0} - \epsilon \sum_{i=1}^{r} n_i \frac{\partial F}{\partial N_i} + \epsilon \sum_{j=1}^{s} m_j \frac{\partial F}{\partial N'_j} = 0.$$

Dividing by ϵ and calculating the derivatives with the aid of equation (168) (the f's are functions of T only and therefore do not vary during an isothermal transformation), we find, on neglecting all terms proportional to the small quantities N_i/N_0 and N'_j/N_0:

$$0 = -n_0 f_0 - \sum_{i=1}^{r} n_i \left\{f_i + RT + RT \log \frac{N_i}{N_0}\right\} + \sum_{j=1}^{s} m_j \left\{f'_j + RT + RT \log \frac{N'_j}{N_0}\right\}$$

[4] Since the variations in volume of a solution are always very small, it is immaterial whether we consider the equilibrium condition at constant volume or at constant pressure.

or

$$\log \left\{ \frac{\left(\frac{N_1}{N_0}\right)^{n_1}\left(\frac{N_2}{N_0}\right)^{n_2}\cdots\left(\frac{N_r}{N_0}\right)^{n_r}}{\left(\frac{N_1'}{N_0}\right)^{m_1}\left(\frac{N_2'}{N_0}\right)^{m_2}\cdots\left(\frac{N_s'}{N_0}\right)^{m_s}} \right.$$

$$= \frac{\sum_{j=1}^{s} m_j(f_j' + RT) - \sum_{i=1}^{r} n_i(f_i + RT) - n_0 f_0}{RT}.$$

The right-hand side of this equation is a function of T only. If we place it equal to $\log K(T)$, K being a convenient function of the temperature, we finally obtain:

$$\frac{\left(\frac{N_1}{N_0}\right)^{n_1}\cdots\left(\frac{N_r}{N_0}\right)^{n_r}}{\left(\frac{N_1'}{N_0}\right)^{m_1}\cdots\left(\frac{N_s'}{N_0}\right)^{m_s}} = K(T). \tag{169}$$

This equation is the expression of the law of mass action for chemical equilibria in solutions.

The discussion of (169) for the case where the solvent does not take part in the reaction (that is, when $n_0 = 0$ in (167)) is the same as the discussion of the law of mass action for gases (see section 24). It follows, in particular, from equation (169) that if we dilute the solution, the equilibrium is shifted in the direction of increasing dissociation. Of course, in this case we have no simple way of determining the form of $K(T)$, as we did in the case of gases. We know only that $K(T)$ is a function of the temperature.

As a particularly important example of the case for which the solvent participates in the chemical reaction, we consider the reaction:

$$H_2O = H^+ + OH^-, \tag{170}$$

that is, the dissociation of water into hydrogen and hydroxyl ions (the hydrolysis of water). Let $[H^+]$ and $[OH^-]$ be the concentrations of the hydrogen and the hydroxyl ions (numbers of moles per cc.). If we consider a cubic centimeter of water, we have $N_0 = \frac{1}{18}$. Hence, the ratios of the number of moles of $[H^+]$ and $[OH^-]$ to the number of moles of water are, respectively, $18[H^+]$ and $18[OH^-]$. Applying equation (169) to the reaction (170), we thus find:

$$\frac{1}{18^2[H^+][OH^-]} = K(T),$$

or

$$[H^+][OH^-] = \frac{1}{18^2 K(T)} = K'(T), \tag{171}$$

where $K'(T)$ is a new function of the temperature only.

We see from this equation that the product of the concentrations of the hydrogen and the hydroxyl ions in water is a constant when the temperature is constant.[5] At room temperature, this product is approximately equal to 10^{-14} when the concentrations are expressed in moles per liter; that is,

$$[H^+][OH^-] = 10^{-14}. \tag{172}$$

In pure water, the concentrations of H^+ and OH^- are equal, so that for this case we have from (172):

$$[H^+] = [OH^-] = 10^{-7}.$$

If we add some acid to the water, there is an increase of $[H^+]$, and, since the product (172) must remain constant, a corresponding decrease of $[OH^-]$.

The opposite occurs if a base is added to the water. It is usual to indicate the acidity of a water solution by the symbol:

$$\text{pH} = -\text{Log}\,[H^+]. \tag{173}$$

(Log stands for the logarithm to the base 10; $[H^+]$ is expressed as before in moles per liter.) Thus, pH = 7 means a

[5] From the law of mass action applied to the reaction (171), one would expect the ratio $[H^+][OH^-]/[H_2O]$ to be a function of T only. Since the denominator is practically constant, however, the numerator also must be a function of T only in accordance with equation (171). We see thus that (171) is essentially equivalent to the law of mass action in its usual form.

neutral reaction; pH < 7 indicates acidity; and pH > 7 indicates a basic reaction.

The above discussion of chemical equilibria in solutions is incomplete, since no account has been taken of the electrostatic forces between ions. It has been shown by Debye and Hückel that such forces are often of importance and may affect the chemical reaction considerably. A discussion of this point, however, lies beyond the scope of this book.

28. The distribution of a solute between two phases. Let A and B be two immiscible liquids (as, for example, water and ethyl ether) in contact. Let C be a third substance soluble both in A and in B. If we dissolve a certain amount of C in the liquid A, the substance C diffuses through the surface that separates A and B; and after a short time, C will be in solution in both liquids. The concentration of C in the liquid B will continue to increase, and the concentration of C in A will decrease until equilibrium is reached between the two solutions.

Let N_A and N_B be the numbers of moles of the two solvents A and B, and let N_1 and N_1' be the numbers of moles of the solute C dissolved in A and B, respectively. The thermodynamic potential, Φ, of our system will be the sum of the potentials of the two solutions.

We have first a solution of N_1 moles of C dissolved in N_A moles of the liquid A. The thermodynamic potential at constant pressure of this solution is, according to (161):

$$\Phi_A = N_A\{f_A(T) + pv_A(T)\} + N_1\{f_1(T) + pv_1(T)\} + RTN_1 \log \frac{N_1}{N_A}, \quad (174)$$

where f_A, f_1, v_A, and v_1 correspond to f_0, f_1, v_0, and v_1 of the general formula (161).

Second, we have a solution which contains N_B moles of the solvent B and N_1' moles of the solute C. Its thermodynamic potential is given by:

$$\Phi_B = N_B\{f_B(T) + pv_B(T)\} + N_1'\{f_1'(T) + pv_1'(T)\} + RTN_1' \log \frac{N_1'}{N_B}, \quad (175)$$

where the quantities f_B, f_1', v_B, and v_1' correspond to f_0, f_1, v_0, and v_1 of (161).

The thermodynamic potential Φ of the complete system is:

$$\Phi = \Phi_A + \Phi_B . \tag{176}$$

For a given temperature and pressure, the equilibrium condition is that Φ be a minimum.

We consider an infinitesimal transformation of our system as a result of which an amount dN_1 of C passes from the liquid B into the liquid A. N_1 and N_1' will change by amounts dN_1 and $-dN_1$, respectively, and the variation in Φ will be given by:

$$dN_1 \frac{\partial \Phi}{\partial N_1} - dN_1 \frac{\partial \Phi}{\partial N_1'} .$$

If Φ is to be a minimum, this expression must vanish. Dividing by dN, we thus obtain the equation:

$$\frac{\partial \Phi}{\partial N_1} = \frac{\partial \Phi}{\partial N_1'} . \tag{177}$$

Using (176), (175), and (174), we obtain the equilibrium condition:

$$f_1(T) + pv_1(T) + RT \log \frac{N_1}{N_A} + RT$$
$$= f_1'(T) + pv_1'(T) + RT \log \frac{N_1'}{N_B} + RT,$$

or

$$\frac{\dfrac{N_1}{N_A}}{\dfrac{N_1'}{N_B}} = e^{\frac{f_1'(T) - f_1(T) + p[v_1'(T) - v_1(T)]}{RT}} = K(T, p), \tag{178}$$

where the function $K(T, p)$ depends only on the temperature and pressure and not on the concentrations.

Equation (178) expresses the following law:

When two dilute solutions of the same solute in two different immiscible solvents are in contact and in equilibrium, the ratio of the concentrations of the two solutions at a given temperature and pressure is constant.

A problem analogous to the preceding one is the following:
A solution of a gas dissolved in a liquid is in contact with the gas itself; to find the relationship between the pressure of the gas and the concentration of the solution for which the system is in equilibrium at a given temperature.

Let N_0 and N_1 be the numbers of moles of the liquid solvent and the gaseous solute in the solution, respectively; and let N_1' be the number of moles of gas in the gaseous phase. Since variations in volume of the solution are practically negligible as compared with variations in volume of the gaseous phase, we can neglect the term pV in the expression for the thermodynamic potential of the solution and identify this potential with the free energy of the solution. According to (159), this is:

$$N_0 f_0(T) + N_1 f_1(T) + RTN_1 \log \frac{N_1}{N_0}. \tag{179}$$

The thermodynamic potential of the gaseous phase is obtained from (125) by multiplying it by the number, N_1', of moles of gas:

$$N_1'[C_pT + W - T(C_p \log T - R \log p + a + R \log R)]. \tag{180}$$

Adding (179) and (180), we obtain the thermodynamic potential Φ of the total system. Just as in the preceding problem, we obtain equation (177) as the condition for equilibrium. Substituting the explicit expressions for the derivatives in (177), we obtain as the condition for equilibrium the following equation:

$$f_1(T) + RT \log \frac{N_1}{N_0} + RT = C_pT$$
$$+ W - T(C_p \log T - R \log p + a + R \log R);$$

or, dividing by RT and passing from logarithms to numbers, we find that:

$$\frac{1}{p}\frac{N_1}{N_0} = e^{\frac{C_pT+W-T(C_p \log T+a+R \log R)-f_1(T)-RT}{RT}}$$

$$= K(T), \tag{181}$$

where $K(T)$ is a function of the temperature alone.

Equation (181) expresses the following law:

The concentration of a solution of a gas dissolved in a liquid at a given temperature is proportional to the pressure of the gas above the solution.

It can be proved in a similar fashion that if there is a mixture of several gases above a liquid, the concentration of each gas in solution is proportional to its partial pressure in the mixture above the liquid. The constant of proportionality in each case depends on the temperature as well as on the nature of the solvent and of the particular gas considered.

29. The vapor pressure, the boiling point, and the freezing point of a solution. The vapor pressure, the boiling point, and the freezing point for a solution are not the same as for the pure solvent. This fact is very important from a practical point of view, because, as we shall show in this section, the changes in the boiling and freezing points, at least for dilute solutions, are proportional to the molecular concentrations of the solutes. The observation of these changes affords, therefore, a very convenient method of determining the molecular concentration of the solution.

We shall assume that the solutes are nonvolatile. In that case, the vapor of the solution will contain only pure vaporized solvent. We shall assume further that, when the solution freezes, only the pure solidified solvent separates out, leaving all the solute still in solution.

We can now show, from very simple considerations, that the vapor pressure for a solution at a given temperature is lower than that for the pure solvent at the same temperature. To this end, we consider the apparatus shown in Figure 21. It consists of a rectangular-shaped tube in which the pure solvent and the solution are separated from each other on the lower side by a semipermeable membrane at B. The levels A and C of the pure solvent and the solution, respectively, will not be at the same height because of the osmotic presure; the level C of the solution will be higher. Since the dissolved substance is nonvolatile, the region in the tube above A and C will be filled with the vapor of the pure solvent only.

We first wait until equilibrium is established; the vapor pressure in the immediate neighborhood of the meniscus A will then be that of a saturated vapor in equilibrium with its liquid phase, and the vapor pressure at C will be that of a saturated vapor in equilibrium with a solution. It is evident that the pressures at A and at C are not equal, since A and C are at different heights in the vapor. Since C lies higher than A, the vapor pressure at C is lower than that at A; that is, the pressure of the vapor above the solution is lower than the vapor pressure above the pure solvent.

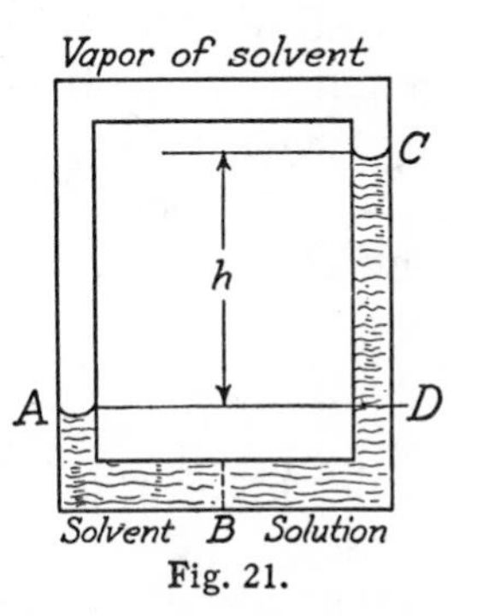

Fig. 21.

To calculate this difference in pressure, Δp, quantitatively, we notice that it is equal to the pressure exerted by a column of vapor of height h. If ρ' is the density of the vapor, and g is the acceleration of gravity, we have:

$$\Delta p = \rho' hg.$$

On the other hand, the pressure exerted by the liquid column CD is equal to the osmotic pressure P of the solution. If ρ is the density of the pure solvent, we have for the osmotic pressure (neglecting the difference between the density of the solution and that of the pure solvent, and also neglecting the density of the vapor as compared to that of the liquid):

$$P = \rho hg.$$

Dividing the first equation by the second, we obtain:

$$\frac{\Delta p}{P} = \frac{\rho'}{\rho},$$

or

$$\Delta p = P\frac{\rho'}{\rho} = P\frac{v_0}{v_0'},$$

where v_0 and v_0' are the volumes occupied by one mole of the pure solvent in the liquid phase and in the vapor phase, respectively (that is, v_0 and v_0' are inversely proportional to ρ and ρ', respectively). Replacing the osmotic pressure P by the expression (165), and assuming, for the sake of simplicity, that there is only one solute present in the solution, we obtain:

$$\Delta p = \frac{RT}{v_0'}\frac{N_1}{N_0}, \tag{182}$$

which is the expression for the difference between the vapor pressure of the solution and that of the pure solvent.

The fact that the vapor pressure for a solution is lower than that for the pure solvent is directly related to the fact that the boiling point of a solution is higher than that of the pure solvent. The reason for this is that the boiling point is the temperature at which the vapor pressure is equal to one atmosphere. Consider a pure solvent at the boiling point; its vapor pressure is equal to one atmosphere. If we now dissolve some substance in this solvent, keeping the temperature constant, the vapor pressure will fall below one atmosphere. Hence, in order to bring the pressure back to its original value of one atmosphere, we must raise the temperature of the solution. With the aid of equation (182) and Clapeyron's equation, one can easily derive an expression for the variation of the boiling point of a solution.

Instead of doing this, however, we shall calculate both the decrease in the vapor pressure and the increase in the boiling point of a solution by a direct method.

We consider a dilute solution composed of N_0 moles of solvent and N_1 moles of a solute in equilibrium with the vapor of the pure solvent. Let N_0' be the number of moles of solvent contained in the vapor phase. From (148), (149), (155), and (121), we obtain for the thermodynamic potential Φ_{sol} of the solution:

$$\Phi_{sol} = N_0 \varphi_0 (T, p) + N_1 \varphi_1 (T, p) + RTN_1 \log \frac{N_1}{N_0},$$

where

$$\varphi_0(T, p) = u_0 - T\sigma_0 + pv_0 , \text{ and } \varphi_1 = u_1 - T\sigma_1 + pv_1 .$$

Let $\varphi_0'(T, p)$ be the thermodynamic potential of one mole of vapor of the solvent. The thermodynamic potential of the N_0' moles of the vapor phase is, then:

$$\Phi_{vap} = N_0' \varphi_0' (T, p);$$

and the thermodynamic potential of the total system is:

$$\Phi = \Phi_{sol} + \Phi_{vap} = N_0 \varphi_0(T, p) + N_1 \varphi_1(T, p) + RTN_1 \log \frac{N_1}{N_0} + N_0' \varphi_0'(T, p). \quad (183)$$

The equilibrium condition is that Φ be a minimum at constant temperature and pressure. We must therefore have $d\Phi = 0$ for an infinitesimal, isothermal, isobaric transformation. If dN_0 moles of the solvent are transferred from the vapor phase to the solution as a result of such a transformation (that is, if N_0 and N_0' vary by the amounts dN_0 and $-dN_0$, respectively), then we must have:

$$d\Phi = dN_0 \frac{\partial \Phi}{\partial N_0} - dN_0 \frac{\partial \Phi}{\partial N_0'} = 0,$$

or

$$\frac{\partial \Phi}{\partial N_0} = \frac{\partial \Phi}{\partial N_0'}.$$

Replacing the derivatives in this equation by their explicit expressions as calculated from (183), we obtain:

$$\varphi_0(T, p) - RT \frac{N_1}{N_0} = \varphi_0'(T, p),$$

or

$$\varphi_0(T, p) - \varphi_0'(T, p) = RT \frac{N_1}{N_0}. \tag{184}$$

This equation expresses the relationship between the temperature and the vapor pressure of our solution.

Let p_0 be the pressure of the saturated vapor of the pure solvent at the temperature T. T and p_0 will satisfy equation (184) if we place $N_1 = 0$ in that equation, because in that case no solute is present. Thus:

$$\varphi_0(T, p_0) - \varphi_0'(T, p_0) = 0. \tag{185}$$

When N_1 moles of solute are dissolved in the solvent, the pressure p of the vapor becomes:

$$p = p_0 + \Delta p,$$

where Δp is a small quantity. Expanding the left-hand side of (184), in powers of Δp up to terms of the first order, we find that:

$$\begin{aligned} RT \frac{N_1}{N_0} &= \varphi_0(T, p_0) - \varphi_0'(T, p_0) + \Delta p \left\{ \frac{\partial \varphi_0(T, p_0)}{\partial p_0} - \frac{\partial \varphi_0'(T, p_0)}{\partial p_0} \right\} \\ &= \Delta p \left\{ \frac{\partial \varphi_0(T, p_0)}{\partial p_0} - \frac{\partial \varphi_0'(T, p_0)}{\partial p_0} \right\}. \end{aligned} \tag{186}$$

Since φ_0 is the thermodynamic potential of one mole of pure solvent, we obtain from (123):

$$\frac{\partial \varphi_0(T, p_0)}{\partial p_0} = v_0,$$

where v_0 is the volume of one mole of solvent; and, similarly,

$$\frac{\partial \varphi_0'(T, p_0)}{\partial p_0} = v_0',$$

where v_0' is the volume of one mole of vapor of the pure solvent. Substituting these expressions in (186), we have:

$$\Delta p = -\frac{RT}{v_0' - v_0}\frac{N_1}{N_0}. \tag{187}$$

Since the volume, v_0', of one mole of vapor is larger than the volume, v_0, of one mole of liquid solvent, Δp is negative; this means that the pressure of the vapor of the solution is lower than that of the pure solvent. If v_0 is negligible as compared to v_0', which we assumed to be the case in the derivation of equation (182), equation (187) becomes identical with (182). (The minus sign means that the vapor pressure of the solution is lower than that of the pure solvent.)

We have deduced the expression for the decrease in the vapor pressure from equation (184). With the aid of the same equation and by a method analogous to the one just used, we can also calculate the change in the boiling point of a solution.

We consider a solution whose temperature is such that the pressure p of its vapor is equal to one atmosphere. Let T_0 be the boiling point of the pure solvent and $T = T_0 + \Delta T$ the boiling point of the solution. Since the vapor pressure at the boiling point is equal to the atmospheric pressure, p, it follows that the vapor pressure of the pure solvent at the temperature T_0 is equal to p. Since $N_1 = 0$ for the pure solvent, we find, with the aid of (184), that:

$$\varphi_0(T_0, p) - \varphi_0'(T_0, p) = 0. \tag{188}$$

Applying (184) to the solution, we obtain:

$$\varphi_0(T_0 + \Delta T, p) - \varphi_0'(T_0 + \Delta T, p) = RT\frac{N_1}{N_0}.$$

Developing the left-hand side of the preceding equation in powers of ΔT, and dropping all terms above the first, we obtain, with the aid of (188), the following equation:

$$\Delta T\left\{\frac{\partial \varphi_0(T_0, p)}{\partial T_0} - \frac{\partial \varphi_0'(T_0, p)}{\partial T_0}\right\} = RT_0\frac{N_1}{N_0}.$$

From (124) we have:

$$\frac{\partial \varphi_0(T_0, p)}{\partial T_0} = -\sigma_0; \qquad \frac{\partial \varphi_0'(T_0, p)}{\partial T_0} = -\sigma_0',$$

where σ_0 and σ_0' are the entropies of one mole of solvent in the liquid and vapor phases, respectively. From the preceding two equations, we now obtain:

$$\Delta T\{\sigma_0' - \sigma_0\} = RT_0 \frac{N_1}{N_0}. \tag{189}$$

Let Λ be the heat of vaporization of one mole of solvent. If we permit one mole of the solvent to vaporize at the boiling point, T_0, the amount of heat absorbed is Λ, and $\dfrac{\Lambda}{T_0}$ is the change in entropy. Hence,

$$\sigma_0' - \sigma_0 = \frac{\Lambda}{T_0}.$$

Substituting this in equation (189), we obtain:

$$\Delta T = \frac{RT_0^2}{\Lambda} \frac{N_1}{N_0}. \tag{190}$$

This is the expression for the difference between the boiling point of the solution and the boiling point of the pure solvent. Since $\Delta T > 0$, the boiling point of the solution is higher than that of the pure solvent. We see also from the equation that the change in the boiling point is proportional to the molecular concentration of the solution.

As an example, we shall apply the above equation to a normal solution of some substance in water. For such a solution, we have:

$$N_1 = 1; \qquad N_0 = \frac{1000}{18}; \qquad \Lambda = 540 \times 18 \text{ calories};$$

$$R = 1.986 \text{ calories}; \qquad T_0 = 373.1°K.$$

(We can express both R and Λ in calories in equation (190) because their ratio is obviously dimensionless.) Substituting these values in equation (190), we find that:

$$\Delta T = 0.51 \text{ degrees.}$$

The same formula (190) can also be used to calculate the change in the freezing point of a solution. The only difference is that, instead of having a vapor phase, we have a solid phase. Λ in that case represents the heat absorbed by one mole of the solvent in passing isothermally from the liquid to the solid state at the freezing point. This heat is negative and equal to $-\Lambda'$, where Λ' is the heat of fusion of one mole of the solvent. For the case of freezing, (190) becomes, therefore,

$$\Delta T = -\frac{RT_0^2}{\Lambda'}\frac{N_1}{N_0}. \tag{191}$$

From this equation we see that the freezing point of a solution is lower than that of the pure solvent; the decrease is proportional to the molecular concentration of the solution.

In the case of a normal solution in water, for which

$$N_1 = 1; \quad N_0 = \frac{1000}{18}; \quad \Lambda' = 80 \times 18 \text{ calories};$$

$$R = 1.986 \text{ calories}; \quad T_0 = 273.1°,$$

we find that:

$$\Delta T = -1.85 \text{ degrees.}$$

It should be noticed that in all these formulae N_1 represents the actual number of moles of substance present in the solution. For electrolytic solutions, therefore, each ion must be considered as an independent molecule. Thus, for the case of very strong electrolytes (having a high degree of dissociation), N_1 is obtained by multiplying the number of moles of solute by the number of ions into which a single molecule of the solute dissociates when in solution.

Problems

1. Calculate the osmotic pressure and the variation in the boiling and freezing points of a solution containing 30 grams of NaCl per liter of water.

2. A solution of sugar ($C_6H_{12}O_6$) in water and a solution of NaCl in water have the same volume and the same osmotic pressure. Find the ratio of the weights of sugar and of sodium chloride.

3. Discuss with the aid of the phase rule the equilibrium of a solution and the vapor of the solvent.

4. The concentration of a saturated solution (the ratio of the number of moles of the solute to the number of moles of the solvent) is a function of the temperature. Express the logarithmic derivative of this function in terms of the temperature and the heat of solution. (Assume that the laws of dilute solutions can be applied also to the saturated solution. The formula can be obtained by applying a method analogous to that used for deriving Clapeyron's equation.)

CHAPTER VIII

The Entropy Constant

30. The Nernst theorem. We have already seen that the definition of the entropy given by (68):

$$S(A) = \int_O^A \frac{dQ}{T},$$

where O is an arbitrarily chosen initial state, is incomplete because the arbitrariness in the choice of the initial state introduces an undetermined additive constant in the definition. As long as we deal only with differences of the entropy, this incompleteness is of no consequence. We have already found, however, that cases arise (for example, in dealing with gaseous equilibria, Chapter VI) for which the knowledge of this constant becomes important. In this chapter we shall introduce and discuss a principle that will enable us to determine the additive constant appearing in the definition of the entropy. This principle, which was discovered by Nernst, is often referred to as the *third law of thermodynamics* or as *Nernst's theorem.*

In the form in which it was originally stated by Nernst, this theorem applied only to condensed systems, but it has since then been extended to apply to gaseous systems also. We may state this theorem in the following form:

The entropy of every system at absolute zero can always be taken equal to zero.

Since we have defined only differences of entropy between any two states of a system, the above statement of Nernst's theorem must be interpreted physically as meaning that all possible states of a system at the temperature $T = 0$ have the same entropy. It is therefore obviously convenient to choose one of the states of the system at $T = 0$ as the standard state O introduced in section 12; this will permit us to set the entropy of the standard state equal to zero.

The entropy of any state A of the system is now defined, including the additive constant, by the integral:

$$S(A) = \int_{T=0}^{A} \frac{dQ}{T}, \tag{192}$$

where the integral is taken along a reversible transformation from any state at $T = 0$ (lower limit) to the state A.

In this book we shall assume Nernst's theorem as a postulate; a few words concerning its theoretical basis, however, will serve to demonstrate its plausibility.

We have seen that a thermodynamical state of a system is not a sharply defined state of the system, because it corresponds to a large number of dynamical states. This consideration led to the Boltzmann relation (75):

$$S = k \log \pi,$$

where π is called the probability of the state. Strictly speaking, π is not the probability of the state, but is actually the number of dynamical states that correspond to the given thermodynamical state. This seems at first sight to give rise to a serious difficulty, since a given thermodynamical state corresponds to an infinite number of dynamical states. This difficulty is avoided in classical statistical mechanics by the following device:

The dynamical states of a system form an ∞^{2f} array, where f is the number of degrees of freedom of the system; each state can therefore be represented by a point in a $2f$-dimensional space, which is called the *phase space of the system.* Instead of an exact representation of the dynamical state, however, which could be given by designating the precise position in the phase space of the point representing the state, the following approximate representation is introduced:

The phase space is divided into a number of very small cells all of which have the same hyper-volume τ; the state is then characterized by specifying the cell to which the point representing the state belongs. Thus, states whose representative points all lie in the same cell are not considered as being different. This representation of the state of a system would evidently become exact if the cells were made infinitesimal.

The cell representation of the dynamical states of a system introduces a discontinuity in the concept of the state of a system which enables us to calculate π by the methods of combinatory analysis, and, hence, with the aid of the Boltzmann relation, to give a statistical definition of the entropy. It should be noticed, however, that the value of π, and therefore the value of the entropy also, depends on the arbitrarily chosen size of the cells; indeed, one finds that, if the volume of the cells is made vanishingly small, both π and S become infinite. It can be shown, however, that if we change τ, π is altered by a factor. But from the Boltzmann relation, $S = k \log \pi$, it follows that an undetermined factor in π gives rise to an undetermined additive constant in S. We see from the foregoing considerations that the classical statistical mechanics cannot lead to a determination of the entropy constant.

The arbitrariness associated with π, and therefore with the entropy also, in the classical picture can be removed by making use of the principles of the quantum theory. The reason for this is that the quantum theory introduces a discontinuity quite naturally into the definition of the dynamical state of a system (the discrete quantum states) without having to make use of the arbitrary division of the phase space into cells. It can be shown that this discontinuity is equivalent, for statistical purposes, to the division of the phase space into cells having a hyper-volume equal to h^f, where h is Planck's constant ($h = 6.55 \times 10^{-27}$ cm.^{2} gm. sec.$^{-1}$) and f is the number of degrees of freedom of the system. We may note here, without entering into the details, which lie outside the scope of this book, that in a statistical theory based consistently on the quantum theory all indeterminacy in the definition of π, and therefore in the definition of the entropy also, disappears.

According to the Boltzmann relation, the value of π which corresponds to $S = 0$ is $\pi = 1$. Statistically interpreted, therefore, Nernst's theorem states that *to the thermodynamical state of a system at absolute zero there corresponds only one dynamical state, namely, the dynamical state of lowest energy compatible with the given crystalline structure or state of aggregation of the system.*

The only circumstances under which Nernst's theorem might be in error are those for which there exist many dynamical states of lowest energy. But even in this case, the number of such states must be enormously large[1] if deviations from the theorem are to be appreciable. Although it is not theoretically impossible to conceive of such a system, it seems extremely unlikely that such systems actually exist in nature. We may therefore assume that Nernst's theorem is generally valid.

We shall now develop some of the consequences of Nernst's theorem.

31. Nernst's theorem applied to solids. We consider a solid body which is heated (at constant pressure, for example) until its temperature increases from the absolute zero to a certain value, T. Let $C(T)$ be its thermal capacity (at constant pressure) when its temperature is T. Then, if the temperature changes by an amount dT, the body will absorb an amount of heat $dQ = C(T)dT$. The entropy of the body at the temperature T is therefore given (see equation (192)) by:

$$S = \int_0^T \frac{C(T)}{T}\, dT. \tag{193}$$

We can obtain the first consequence of Nernst's theorem from equation (193): we observe that if the thermal capacity, $C(0)$, at absolute zero were different from zero,

[1] Of the order of e^N, where N is the number of molecules in the system.

the integral (193) would diverge at the lower limit. We must therefore have:

$$C(0) = 0. \tag{194}$$

This result is in agreement with the experiments on the specific heats of solids.

We shall limit ourselves here, for the sake of simplicity, to the consideration of solid chemical elements, and perform the calculations for one gram atom of the element. Figure 22 is a graphical representation of the general way in which the atomic heats of solids depend on the temperature as found empirically. One can see from the figure that the atomic heat actually vanishes at absolute zero. At higher temperatures, $C(T)$ approaches a limiting value which is very nearly the same for all solid elements and which lies very close to the value $3R$. Since this limiting value is practically attained at room temperature, this result is an expression of the well-known law of Dulong and Petit, which can be stated as follows:

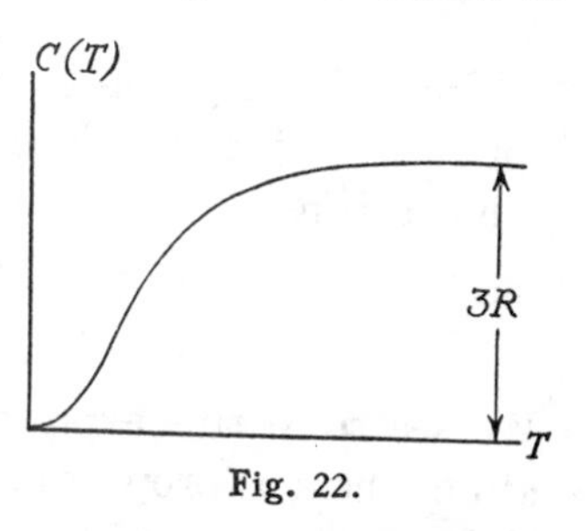

Fig. 22.

All solid elements at room temperature have the same atomic heat, which is equal to 3R (*that is, the product: specific heat* × *atomic weight is the same for all solids and is equal to* 3R).

A theoretical formula for the specific heats of solid elements, which is in very good agreement with experiment, was derived by Debye on the basis of the quantum theory. The Debye expression can be written in the form:

$$C(T) = 3RD\left(\frac{T}{\Theta}\right), \tag{195}$$

where Θ is a characteristic constant of the substance, which has the dimensions of a temperature; it is called the *Debye temperature.* D represents the following function:

$$D(\xi) = 12\xi^3 \int_0^{\frac{1}{\xi}} \frac{x^3\,dx}{e^x - 1} - \frac{\frac{3}{\xi}}{e^{1/\xi} - 1}. \tag{196}$$

Since $D(\xi)$ approaches the limit 1 for large values of ξ, it follows from (195) that the atomic heat for high temperatures tends to the limit $3R$, as required by the law of Dulong and Petit.

For small values of ξ, we may replace the upper limit of the integral in (196) by infinity, and we may neglect the second term in that expression because that term becomes an infinitesimal of a very high order for infinitesimal values of ξ. For $\xi \to 0$, we therefore obtain:

$$D(\xi) \to 12\xi^3 \int_0^{\infty} \frac{x^3\,dx}{e^x - 1} = \frac{4\pi^4}{5}\xi^3. \tag{197}$$

From this asymptotic expression for $D(\xi)$, we obtain the following expression for the atomic heat in the limit of low temperatures:

$$C(T) = \frac{12\pi^4}{5}\frac{R}{\Theta^3}T^3 + \cdots. \tag{198}$$

We see from this expression that at low temperatures the atomic heat is proportional to the cube of the temperature. This consequence of the Debye theory is in good agreement with experiment.

Using the Debye formula, we can calculate the entropy of a gram atom of our substance by substituting (195) in (193). On doing this, we find that:

$$S = \int_0^T \frac{C(T)}{T}\,dT = 3R\int_0^T D\left(\frac{T}{\Theta}\right)\frac{dT}{T} = 3R\int_0^{\frac{T}{\Theta}} D(\xi)\,\frac{d\xi}{\xi}. \tag{199}$$

Replacing $D(\xi)$ in (199) by its explicit expression, we find that[2]:

[2] The following integral formulae are used:

$$\int_0^{\omega} D(\xi)\,\frac{d\xi}{\xi} = 12\int_0^{\omega} \xi^2\,d\xi \int_0^{\frac{1}{\xi}} \frac{x^3\,dx}{e^x - 1} - 3\int_0^{\omega} \frac{\frac{d\xi}{\xi^2}}{e^{1/\xi} - 1},$$

$$S = 3R\left\{4\frac{T^3}{\Theta^3}\int_0^{\frac{\Theta}{T}}\frac{x^3\,dx}{e^x - 1} - \log\left(1 - e^{-\frac{\Theta}{T}}\right)\right\}$$

$$= 3R\log T + 4R - 3R\log\Theta + \cdots, \tag{200}$$

where the last formula is valid for $T \gg \Theta$, that is, in the range of temperatures for which the law of Dulong and Petit holds.

With the aid of Nernst's theorem, we shall now discuss the transformation of a solid from one crystalline form to another. As an example, we shall consider the transformation from grey to white tin. Grey tin is the stable form at low temperatures and white tin is stable at high temperatures. The transition temperature, T_0, is equal to 19°C or 292°K.

The transformation of tin from one of these allotropic forms to the other is analogous in many respects to the melting of a solid. Thus, for example, a certain amount of heat is absorbed by the tin in passing from the grey to the white form. This heat of transformation, Q, is equal to 535 calories per gram-atom at the transition temperature.

Although grey tin is the stable form below the transition temperature, white tin can exist in a labile form down to the lowest temperatures. It is therefore possible to measure the specific heats of both grey and white tin all the way from the lowest temperatures to the transition temperature. The atomic heats of the two forms are not equal; the atomic

---

or, interchanging the order of integration in the double integral, and introducing $1/\xi$ as a new variable in the second integral, we obtain:

$$\int_0^\omega D(\xi)\frac{d\xi}{\xi} = 12\int_0^{\frac{1}{\omega}}\frac{x^3\,dx}{e^x - 1}\int_0^\omega \xi^2\,d\xi + 12\int_{\frac{1}{\omega}}^\infty \frac{x^3\,dx}{e^x - 1}\int_0^{\frac{1}{x}}\xi^2\,d\xi - 3\int_{\frac{1}{\omega}}^\infty \frac{dx}{e^x - 1}$$

$$= 4\omega^3\int_0^{\frac{1}{\omega}}\frac{x^3\,dx}{e^x - 1} - \log\left(1 - e^{-\frac{1}{\omega}}\right).$$

For large values of ω, we obtain the following asymptotic expression:

$$\int_0^\omega D(\xi)\frac{d\xi}{\xi} = \frac{4}{3} + \log\omega + \cdots.$$

heat of grey tin at a given temperature is less than that of white tin at the same temperature.

The transformation from white to grey tin is nonreversible at temperatures below the transition temperature (since the grey form is stable below the transition temperature, a spontaneous transformation can occur only from the white to the grey form). At the transition temperature, however, the transformation between the two forms is reversible.

If $S_1(T_0)$ and $S_2(T_0)$ are the entropies at the transition temperature of one gram-atom of grey and white tin, respectively, then, applying (69) to the reversible, isothermal transformation from grey to white tin, we obtain:

$$S_2(T_0) - S_1(T_0) = \int_{\text{grey}}^{\text{white}} \frac{dQ}{T_0} = \frac{Q}{T_0}. \tag{201}$$

If we indicate the atomic heats of grey and white tin by $C_1(T)$ and $C_2(T)$, respectively, we can express $S_1(T_0)$ and $S_2(T_0)$, with the aid of equation (193), as follows:

$$S_1(T_0) = \int_0^{T_0} \frac{C_1(T)}{T}\,dT; \qquad S_2(T_0) = \int_0^{T_0} \frac{C_2(T)}{T}\,dT. \tag{202}$$

We thus obtain from (201) the equation:

$$Q = T_0\left\{\int_0^{T_0} \frac{C_2(T)}{T}\,dT - \int_0^{T_0} \frac{C_1(T)}{T}\,dT\right\}, \tag{203}$$

which expresses the heat of transformation, Q, of the process in terms of the transition temperature T_0 and the atomic heats of the two forms of tin.

In order to test the validity of equation (203), we shall perform the two integrations indicated numerically. The results of the numerical integrations are:

$$\int_0^{T_0} \frac{C_2(T)}{T}\,dT = 12.30\,\frac{\text{cal.}}{\text{degrees}};$$

$$\int_0^{T_0} \frac{C_1(T)}{T}\,dT = 10.53\,\frac{\text{cal.}}{\text{degrees}}.$$

Since $T_0 = 292$, we obtain from (203):

$$Q = 292\,(12.30 - 10.53) = 517 \text{ cal.}$$

The good agreement between this value and the experimental value, $Q = 535$ calories, can be taken as strong evidence in support of Nernst's theorem. The small difference between the two values can be accounted for by the experimental errors.

32. The entropy constant of gases. In section 14 we calculated the entropy of one mole of an ideal gas (see equation (86)) and found that:

$$S = C_V \log T + R \log V + a.$$

The undetermined additive constant a which appears in this expression is called the *entropy constant* of the gas.

If we could apply Nernst's theorem directly to the formula (86) for the entropy, we could hope to determine a from the condition that the entropy S must vanish at $T = 0$. If we attempt to do this, however, we see that the term $C_V \log T$ on the right-hand side of (86) becomes infinite, and we obtain an infinite value for the entropy constant.

The reason for this apparent failure of Nernst's theorem for ideal gases is that we assumed, as one of the properties of an ideal gas, that the specific heat C_V is a constant; we have already shown (at the beginning of the preceding section) that this is incompatible with Nernst's theorem.

One way out of this difficulty could be sought in the fact that no real substance behaves even approximately like an ideal gas in the neighborhood of absolute zero: all gases condense for sufficiently low temperatures. It is therefore physically not permissible to apply (86) to a gas in the neighborhood of $T = 0$.

But quite apart from this consideration, it follows from quantum mechanics that, even for an ideal gas (defined as a gas whose molecules have a negligible size and do not exert forces on each other), the specific heat at very low temperatures decreases in such a way as to vanish in the neighborhood of $T = 0$. Thus, even for an ideal gas as defined above, (86) can be applied only if the temperature is not too low.

By statistical methods and also by a straightforward application of Nernst's theorem, it is possible to calculate the entropy of an ideal gas for all temperatures. In the limit of high temperatures, the entropy takes the form (86), with the constant a, instead of being undetermined, expressed as a function of the molecular weight and the other molecular constants of the gas.

The simplest case is that of a monatomic gas, for which the entropy of one mole is given by:

$$S = R\left\{\frac{3}{2}\log T + \log V + \log\frac{(2\pi MR)^{\frac{3}{2}}\omega e^{\frac{5}{2}}}{h^3 A^4}\right\}, \qquad (204)$$

where M is the atomic weight; h is Planck's constant ($= 6.55 \times 10^{-27}$ C. G. S. units); A is Avogadro's number ($= 6.03 \times 10^{23}$); and ω is a small integer that is called the *statistical weight of the ground state of the atom.* The value of ω for different atoms is obtained from the quantum theory; we shall give the value of ω for all the examples considered here. e is the base of the natural logarithms.

Formula (204) was first obtained by Tetrode and Sackur. In order to show that (204) can be put in the form (86), we must take (34) into account. On doing this, we obtain for the entropy constant of one mole of a monatomic gas the expression:

$$\begin{aligned} a &= R\log\frac{(2\pi MR)^{\frac{3}{2}}\omega e^{\frac{5}{2}}}{h^3 A^4} \\ &= R\left(-5.65 + \frac{3}{2}\log M + \log\omega\right). \end{aligned} \qquad (205)$$

We can also write the entropy of an ideal monatomic gas in a form corresponding to (87):

$$S = R\left\{\frac{5}{2}\log T - \log p + \log\frac{(2\pi M)^{\frac{3}{2}} R^{\frac{5}{2}}\omega e^{\frac{5}{2}}}{h^3 A^4}\right\}. \qquad (206)$$

We cannot give a proof of these formulae in this book; we shall therefore limit ourselves to some examples showing the applications of these formulae. As a first example, we shall consider the problem of calculating the vapor pressure for a solid monatomic substance.

Let p be the vapor pressure of the substance at the temperature T. Keeping the temperature (and the pressure) constant, we vaporize one mole of the substance by increasing the volume very slowly. During this process, the body absorbs from the environment an amount of heat, Λ, equal to the heat of vaporization (per mole, not per gram). Since the vaporization of the one mole of substance occurs reversibly, the change in entropy during the transformation is:

$$S_{\text{vapor}} - S_{\text{solid}} = \frac{\Lambda}{T}.$$

Using the approximate expression (200) for the entropy of the solid and the formula (206) for the entropy of the vapor, we obtain:

$$R\left\{\frac{5}{2}\log T - \log p + \log \frac{(2\pi M)^{\frac{3}{2}} R^{\frac{5}{2}} \omega e^{\frac{5}{2}}}{h^3 A^4}\right\} - 3R\log T - 4R + 3R\log\Theta = \frac{\Lambda}{T},$$

or, passing from logarithms to numbers,

$$p = \frac{(2\pi M)^{\frac{3}{2}} R^{\frac{5}{2}} \omega \Theta^3}{e^{\frac{3}{2}} h^3 A^4} \frac{1}{\sqrt{T}} e^{-\frac{\Lambda}{RT}} \tag{207}$$

This formula should be compared with (98), which was obtained from Clapeyron's equation. The factor $1/\sqrt{T}$ in (207) arises from our having taken into account the dependence of the heat of vaporization on the temperature. We see that the factor of proportionality, which remained undetermined in (98), has now been completely determined in (207) by the use of Nernst's theorem and the Sackur-Tetrode formula for the entropy of a gas.

Since in many cases we have to deal with the vaporization of a liquid and not of a solid, (207) cannot be used in general. As an example of the vaporization of a liquid, we shall consider the vaporization of one mole of mercury, because this element has a monatomic vapor.

The boiling point of mercury is 630°K. This means that the vapor pressure of saturated mercury vapor at 630°K is equal to one atmosphere.

We shall now calculate the entropy of one mole of mercury at $T = 630$°K and $p = 1$ atmosphere by two different methods and compare the two results.

Method 1. The Sackur-Tetrode formula (206) applied to our case (the atomic weight of mercury is 200.6) gives:

$$S = 191 \times 10^7.$$

Method 2. We start with one mole of solid mercury at absolute zero. Its entropy, according to Nernst's theorem, is zero. We then heat the one mole of mercury, keeping the pressure equal to one atmosphere, until its temperature has reached the melting point, $T_{\text{melting}} = 234.2$°K. During this process the entropy of the mercury increases; its value for $T = 234.2$°K can be calculated with the aid of (193):

$$S_{\text{solid}}\ (243.2) = \int_0^{243.2} \frac{C(T)}{T}\, dT,$$

where $C(T)$ is the atomic heat at constant pressure of mercury. The above integral can be calculated numerically by using the experimentally determined values of $C(T)$. On doing this, we obtain:

$$S_{\text{solid}}(243.2) = 59.9 \times 10^7.$$

We now let the mole of mercury melt at atmospheric pressure. During this process, the body absorbs reversibly an amount of heat equal to the heat of fusion for one mole of mercury (2330×10^7 ergs/mole). The change in entropy resulting from this is therefore obtained by dividing the heat of fusion by the melting point; that is, the change in entropy is equal to $2330 \times 10^7/243.2 = 9.9 \times 10^7$. The total entropy of the mole of mercury is now:

$$S_{\text{liquid}}(243.2) = 59.9 \times 10^7 + 9.9 \times 10^7 = 69.8 \times 10^7.$$

Next we heat the liquid mercury and raise its temperature from the melting point to the boiling point. During this process, the entropy changes by the amount:

$$S_{\text{liquid}}\,(630°) - S_{\text{liquid}}\,(243.2°) = \int_{243.2}^{630} \frac{C_l(T)}{T}\,dT,$$

where $C_l(T)$ is the atomic heat at constant pressure. Using the experimental values of $C_l(T)$, we can evaluate this integral numerically. Its value is 26.2×10^7. Adding this to the value of the entropy of the liquid mercury at the melting point, we find that:

$$S_{\text{liquid}}(630°) = 69.8 \times 10^7 + 26.2 \times 10^7 = 96.0 \times 10^7.$$

We finally permit the mole of liquid mercury to vaporize at atmospheric pressure. As a result of this, the mercury at the temperature $T = 630°$ absorbs an amount of heat equal to the heat of vaporization of one mole of mercury ($59{,}300 \times 10^7$ ergs/mole). The change in entropy is therefore equal to $59{,}300 \times 10^7/630 = 94 \times 10^7$, and we finally obtain for the entropy of the mole of mercury vapor at the boiling temperature:

$$S = 96 \times 10^7 + 94 \times 10^7 = 190 \times 10^7.$$

This is in excellent agreement with the value found directly from the Sackur-Tetrode formula.

The result which we have just obtained may be taken as an experimental proof of the expression for the entropy of a monatomic gas. Similar calculations have been performed for argon and carbon, and in these cases also very satisfactory agreement was found.

33. Thermal ionization of a gas: the thermionic effect. In Chapter VI we established the law of mass action (equation (139)) for chemical equilibria in gaseous systems. The constant coefficient (the factor which does not contain the temperature) on the left-hand side of equation (139) contains the entropy constants of gases that take part in the reaction. The knowledge of the entropy constants enables us, therefore, to calculate this coefficient completely.

Since we gave the expression for the entropy constant of a gas only for monatomic gases, we must choose, as an example, a reaction in which only monatomic gases take part. It is evident that no reaction of this kind can be found in chemistry. We shall therefore consider the following nonchemical process.

When a gas, such, for example, as an alkali vapor, is heated to a very high temperature, some of its atoms become ionized; that is, they lose one of their electrons, and are thus changed into ions. If, for example, we denote by Na, Na^+, and e sodium atoms, sodium ions, and electrons, respectively, the process may be represented by the reaction:

$$\mathrm{Na} \rightleftarrows \mathrm{Na}^+ + \mathrm{e}. \tag{208}$$

It is found that, at any given temperature, this ionization reaction reaches a state of thermal equilibrium which is quite analogous to the chemical equilibrium for ordinary chemical reactions.

In sodium vapor at very high temperatures, we actually have a mixture of three different gases:

Neutral sodium, Na, having a concentration [Na]; sodium ions, Na^+, having a concentration $[Na^+]$; and an electron gas (a gas composed of free electrons), having a concentration [e].

Each of these three substances behaves like a monatomic gas; we may therefore apply the general results, in particular, equation (139), of the theory of chemical equilibria in gaseous systems to the ionization process (208).

Since all the gases in the mixture are monatomic, we must use the first of the expressions (34) for the molecular heats of the gases. The entropy constants can be found with the aid of equation (205); and the statistical weights ω are equal to 2, 1, and 2 for neutral sodium, sodium ions, and electrons, respectively. We place $M = 23$, the atomic weight of sodium, and neglect the very small difference between the masses of sodium atoms and sodium ions, so that we may also place M equal to the atomic weight of the sodium ions. The atomic weight of the electrons (that is, the mass of the electrons divided by $\frac{1}{16}$ of the mass of oxygen) is $M_e = \frac{1}{1830}$. Let us finally denote by W ($= 4.91 \times 10^{-12}$ ergs/mole) the energy needed to ionize all the atoms in one mole of sodium vapor. We have, then,

$$\sum m_j W_j - \sum n_i W_i = W_{\text{ions}} + W_{\text{electrons}} - W_{\text{atoms}} = W.$$

Making all the necessary substitutions in equation (139), we finally obtain, as the condition for thermal equilibrium in the thermal ionization of sodium vapor, the following equation:

$$\frac{[\text{Na}]}{[\text{Na}^+]\,[e]} = \frac{h^3 A^4}{(2\pi M_e R)^{\frac{3}{2}}} T^{-\frac{3}{2}} e^{\frac{W}{RT}}.$$

This formula can be put into a more convenient form as follows: Let x be the degree of ionization, that is, the fraction of atoms that are ionized:

$$x = \frac{[\text{Na}^+]}{[\text{Na}] + [\text{Na}^+]};$$

and let $n = [\text{Na}] + [\text{Na}^+]$ be the total concentration of the sodium (atoms + ions). We have, then,

$$[\text{Na}^+] = nx; \qquad [\text{Na}] = n(1 - x).$$

Since there is obviously one electron present for each sodium ion, we have:

$$[e] = [\text{Na}^+] = nx,$$

and we finally obtain:

$$n \frac{x^2}{1 - x} = \frac{(2\pi M_e R)^{\frac{3}{2}}}{h^3 A^4} T^{\frac{3}{2}} e^{-\frac{W}{RT}}$$

$$= 3.9 \times 10^{-9}\, T^{\frac{3}{2}}\, 10^{-\frac{26,000}{T}}. \qquad \textbf{(209)}$$

The degree of ionization can be calculated from this formula.

Equation (209), which was first derived by M. N. Saha, has found several important applications in the physics of stellar atmospheres.

As a further application of the Sackur-Tetrode formula, we shall obtain the expression for the density of an electron gas which is in equilibrium with a hot metal surface. When a metal is heated to a sufficiently high temperature, it gives off a continuous stream of electrons. If we heat a block of metal containing a cavity, the electrons coming from the metal will fill the cavity until a state of equilibrium is reached, when as many electrons will be reabsorbed per unit time by the metal as are emitted. We propose to calculate the equilibrium concentration of the electrons inside the cavity as a function of the temperature.

Let N be the number of moles of electrons inside the cavity of volume V. The entropy of these electrons is obtained from (204) by multiplying that expression by N and replacing V in it by V/N, since V/N is the volume occupied by one mole of the electron gas. Making use of (34) and (29), we obtain for the energy of the electrons:

$$U = N(\tfrac{3}{2}RT + W),$$

where W is the energy needed to extract one mole of electrons from the metal.

For the free energy of the electron gas, we now obtain the expression:

$$F_{\text{el}} = N(\tfrac{3}{2}RT + W) - NRT\left\{\tfrac{3}{2}\log T + \log\frac{V}{N} + \log\frac{(2\pi M_e R)^{\frac{3}{2}}\, 2e^{\frac{5}{2}}}{h^3 A^4}\right\},$$

where we have put $M_e = \frac{1}{1{,}830}$ = the atomic weight of the electrons, and ω for the electrons = 2.

The free energy F of our complete system is the sum of the previous expression and the free energy F_M of the metal:

$$F = F_M + N\left[\tfrac{3}{2}RT + W - RT\left\{\tfrac{3}{2}\log T + \log V - \log N + \log\frac{2(2\pi M_e R)^{\frac{3}{2}} e^{\frac{5}{2}}}{h^3 A^4}\right\}\right]. \quad (210)$$

The condition for equilibrium is that F be a minimum for a given temperature and volume. Assuming that F_M is independent[3] of N, we thus obtain:

$$0 = \frac{dF}{dN} = \frac{3}{2}RT + W - RT\left\{\frac{3}{2}\log T + \log V - \log N + \log\frac{2(2\pi M_e R)^{\frac{3}{2}} e^{\frac{5}{2}}}{h^3 A^4}\right\} + RT.$$

Passing from logarithms to numbers, we obtain the equation:

$$\frac{N}{V} = \frac{2(2\pi M_e R)^{\frac{3}{2}}}{h^3 A^4} T^{\frac{3}{2}} e^{-\frac{W}{RT}} = 7.89 \times 10^{-9}\, T^{\frac{3}{2}} e^{-\frac{W}{RT}}, \quad (211)$$

which gives, as required, the concentration of the electron gas within the cavity.

Problems

1. Calculate the degree of dissociation of sodium vapor at a temperature of 4,000° K and a pressure of 1 cm. of mercury. (Take into account not only the pressure due to the sodium atoms, but also the contribution of the ions and the electrons.)

2. Find the relation between the Debye temperature Θ and the temperature for which the atomic heat of a solid element is equal to $3R/2$. (Apply graphical or numerical methods.)

---

[3] The experimental basis for this assumption is that the electrons inside a metal do not contribute to the specific heat of the metal; the specific heat is completely accounted for by the motion of the atoms. For a rigorous justification of this assumption, see any treatise on the theory of metals.

Index

R

S

T

U

V

W

Z

A CATALOG OF SELECTED

DOVER BOOKS

IN ALL FIELDS OF INTEREST

A CATALOG OF SELECTED DOVER BOOKS IN ALL FIELDS OF INTEREST

THE BOOK OF BEASTS: Being a Translation from a Latin Bestiary of the Twelfth Century, T. H. White. Wonderful catalog real and fanciful beasts: manticore, griffin, phoenix, amphivius, jaculus, many more. White's witty erudite commentary on scientific, historical aspects. Fascinating glimpse of medieval mind. Illustrated. 296pp. 5⅜ × 8¼. (Available in U.S. only) 24609-4 Pa. $5.95

FRANK LLOYD WRIGHT: ARCHITECTURE AND NATURE With 160 Illustrations, Donald Hoffmann. Profusely illustrated study of influence of nature—especially prairie—on Wright's designs for Fallingwater, Robie House, Guggenheim Museum, other masterpieces. 96pp. 9¼ × 10¾. 25098-9 Pa. $7.95

FRANK LLOYD WRIGHT'S FALLINGWATER, Donald Hoffmann. Wright's famous waterfall house: planning and construction of organic idea. History of site, owners, Wright's personal involvement. Photographs of various stages of building. Preface by Edgar Kaufmann, Jr. 100 illustrations. 112pp. 9¼ × 10.
23671-4 Pa. $7.95

YEARS WITH FRANK LLOYD WRIGHT: Apprentice to Genius, Edgar Tafel. Insightful memoir by a former apprentice presents a revealing portrait of Wright the man, the inspired teacher, the greatest American architect. 372 black-and-white illustrations. Preface. Index. vi + 228pp. 8¼ × 11. 24801-1 Pa. $9.95

THE STORY OF KING ARTHUR AND HIS KNIGHTS, Howard Pyle. Enchanting version of King Arthur fable has delighted generations with imaginative narratives of exciting adventures and unforgettable illustrations by the author. 41 illustrations. xviii + 313pp. 6⅛ × 9¼. 21445-1 Pa. $5.95

THE GODS OF THE EGYPTIANS, E. A. Wallis Budge. Thorough coverage of numerous gods of ancient Egypt by foremost Egyptologist. Information on evolution of cults, rites and gods; the cult of Osiris; the Book of the Dead and its rites; the sacred animals and birds; Heaven and Hell; and more. 956pp. 6⅛ × 9¼.
22055-9, 22056-7 Pa., Two-vol. set $20.00

A THEOLOGICO-POLITICAL TREATISE, Benedict Spinoza. Also contains unfinished *Political Treatise*. Great classic on religious liberty, theory of government on common consent. R. Elwes translation. Total of 421pp. 5⅜ × 8½.
20249-6 Pa. $6.95

INCIDENTS OF TRAVEL IN CENTRAL AMERICA, CHIAPAS, AND YUCATAN, John L. Stephens. Almost single-handed discovery of Maya culture; exploration of ruined cities, monuments, temples; customs of Indians. 115 drawings. 892pp. 5⅜ × 8½. 22404-X, 22405-8 Pa., Two-vol. set $15.90

LOS CAPRICHOS, Francisco Goya. 80 plates of wild, grotesque monsters and caricatures. Prado manuscript included. 183pp. 6⅝ × 9⅝. 22384-1 Pa. $4.95

AUTOBIOGRAPHY: The Story of My Experiments with Truth, Mohandas K. Gandhi. Not hagiography, but Gandhi in his own words. Boyhood, legal studies, purification, the growth of the Satyagraha (nonviolent protest) movement. Critical, inspiring work of the man who freed India. 480pp. 5⅜ × 8½. (Available in U.S. only)
24593-4 Pa. $6.95

HOW TO WRITE, Gertrude Stein. Gertrude Stein claimed anyone could understand her unconventional writing—here are clues to help. Fascinating improvisations, language experiments, explanations illuminate Stein's craft and the art of writing. Total of 414pp. 4⅝ × 6⅝. 23144-5 Pa. $5.95

ADVENTURES AT SEA IN THE GREAT AGE OF SAIL: Five Firsthand Narratives, edited by Elliot Snow. Rare true accounts of exploration, whaling, shipwreck, fierce natives, trade, shipboard life, more. 33 illustrations. Introduction. 353pp. 5⅜ × 8½. 25177-2 Pa. $7.95

THE HERBAL OR GENERAL HISTORY OF PLANTS, John Gerard. Classic descriptions of about 2,850 plants—with over 2,700 illustrations—includes Latin and English names, physical descriptions, varieties, time and place of growth, more. 2,706 illustrations. xlv + 1,678pp. 8½ × 12¼. 23147-X Cloth. $75.00

DOROTHY AND THE WIZARD IN OZ, L. Frank Baum. Dorothy and the Wizard visit the center of the Earth, where people are vegetables, glass houses grow and Oz characters reappear. Classic sequel to *Wizard of Oz*. 256pp. 5⅜ × 8. 24714-7 Pa. $4.95

SONGS OF EXPERIENCE: Facsimile Reproduction with 26 Plates in Full Color, William Blake. This facsimile of Blake's original "Illuminated Book" reproduces 26 full-color plates from a rare 1826 edition. Includes "The Tyger," "London," "Holy Thursday," and other immortal poems. 26 color plates. Printed text of poems. 48pp. 5¼ × 7. 24636-1 Pa. $3.50

SONGS OF INNOCENCE, William Blake. The first and most popular of Blake's famous "Illuminated Books," in a facsimile edition reproducing all 31 brightly colored plates. Additional printed text of each poem. 64pp. 5¼ × 7. 22764-2 Pa. $3.50

PRECIOUS STONES, Max Bauer. Classic, thorough study of diamonds, rubies, emeralds, garnets, etc.: physical character, occurrence, properties, use, similar topics. 20 plates, 8 in color. 94 figures. 659pp. 6⅛ × 9¼. 21910-0, 21911-9 Pa., Two-vol. set $14.90

ENCYCLOPEDIA OF VICTORIAN NEEDLEWORK, S. F. A. Caulfeild and Blanche Saward. Full, precise descriptions of stitches, techniques for dozens of needlecrafts—most exhaustive reference of its kind. Over 800 figures. Total of 679pp. 8⅛ × 11. Two volumes. Vol. 1 22800-2 Pa. $10.95
Vol. 2 22801-0 Pa. $10.95

THE MARVELOUS LAND OF OZ, L. Frank Baum. Second Oz book, the Scarecrow and Tin Woodman are back with hero named Tip, Oz magic. 136 illustrations. 287pp. 5⅜ × 8½. 20692-0 Pa. $5.95

WILD FOWL DECOYS, Joel Barber. Basic book on the subject, by foremost authority and collector. Reveals history of decoy making and rigging, place in American culture, different kinds of decoys, how to make them, and how to use them. 140 plates. 156pp. 7⅞ × 10¾. 20011-6 Pa. $7.95

HISTORY OF LACE, Mrs. Bury Palliser. Definitive, profusely illustrated chronicle of lace from earliest times to late 19th century. Laces of Italy, Greece, England, France, Belgium, etc. Landmark of needlework scholarship. 266 illustrations. 672pp. 6⅛ × 9¼. 24742-2 Pa. $14.95

ILLUSTRATED GUIDE TO SHAKER FURNITURE, Robert Meader. All furniture and appurtenances, with much on unknown local styles. 235 photos. 146pp. 9 × 12. 22819-3 Pa. $7.95

WHALE SHIPS AND WHALING: A Pictorial Survey, George Francis Dow. Over 200 vintage engravings, drawings, photographs of barks, brigs, cutters, other vessels. Also harpoons, lances, whaling guns, many other artifacts. Comprehensive text by foremost authority. 207 black-and-white illustrations. 288pp. 6 × 9. 24808-9 Pa. $8.95

THE BERTRAMS, Anthony Trollope. Powerful portrayal of blind self-will and thwarted ambition includes one of Trollope's most heartrending love stories. 497pp. 5⅜ × 8½. 25119-5 Pa. $8.95

ADVENTURES WITH A HAND LENS, Richard Headstrom. Clearly written guide to observing and studying flowers and grasses, fish scales, moth and insect wings, egg cases, buds, feathers, seeds, leaf scars, moss, molds, ferns, common crystals, etc.—all with an ordinary, inexpensive magnifying glass. 209 exact line drawings aid in your discoveries. 220pp. 5⅜ × 8½. 23330-8 Pa. $3.95

RODIN ON ART AND ARTISTS, Auguste Rodin. Great sculptor's candid, wide-ranging comments on meaning of art; great artists; relation of sculpture to poetry, painting, music; philosophy of life, more. 76 superb black-and-white illustrations of Rodin's sculpture, drawings and prints. 119pp. 8⅜ × 11¼. 24487-3 Pa. $6.95

FIFTY CLASSIC FRENCH FILMS, 1912–1982: A Pictorial Record, Anthony Slide. Memorable stills from Grand Illusion, Beauty and the Beast, Hiroshima, Mon Amour, many more. Credits, plot synopses, reviews, etc. 160pp. 8¼ × 11. 25256-6 Pa. $11.95

THE PRINCIPLES OF PSYCHOLOGY, William James. Famous long course complete, unabridged. Stream of thought, time perception, memory, experimental methods; great work decades ahead of its time. 94 figures. 1,391pp. 5⅜ × 8½. 20381-6, 20382-4 Pa., Two-vol. set $19.90

BODIES IN A BOOKSHOP, R. T. Campbell. Challenging mystery of blackmail and murder with ingenious plot and superbly drawn characters. In the best tradition of British suspense fiction. 192pp. 5⅜ × 8½. 24720-1 Pa. $3.95

CALLAS: PORTRAIT OF A PRIMA DONNA, George Jellinek. Renowned commentator on the musical scene chronicles incredible career and life of the most controversial, fascinating, influential operatic personality of our time. 64 black-and-white photographs. 416pp. 5⅜ × 8¼. 25047-4 Pa. $7.95

GEOMETRY, RELATIVITY AND THE FOURTH DIMENSION, Rudolph Rucker. Exposition of fourth dimension, concepts of relativity as Flatland characters continue adventures. Popular, easily followed yet accurate, profound. 141 illustrations. 133pp. 5⅜ × 8½. 23400-2 Pa. $3.50

HOUSEHOLD STORIES BY THE BROTHERS GRIMM, with pictures by Walter Crane. 53 classic stories—Rumpelstiltskin, Rapunzel, Hansel and Gretel, the Fisherman and his Wife, Snow White, Tom Thumb, Sleeping Beauty, Cinderella, and so much more—lavishly illustrated with original 19th century drawings. 114 illustrations. x + 269pp. 5⅜ × 8½. 21080-4 Pa. $4.50

SUNDIALS, Albert Waugh. Far and away the best, most thorough coverage of ideas, mathematics concerned, types, construction, adjusting anywhere. Over 100 illustrations. 230pp. 5⅜ × 8½. 22947-5 Pa. $4.00

PICTURE HISTORY OF THE NORMANDIE: With 190 Illustrations, Frank O. Braynard. Full story of legendary French ocean liner: Art Deco interiors, design innovations, furnishings, celebrities, maiden voyage, tragic fire, much more. Extensive text. 144pp. 8⅜ × 11¾. 25257-4 Pa. $9.95

THE FIRST AMERICAN COOKBOOK: A Facsimile of "American Cookery," 1796, Amelia Simmons. Facsimile of the first American-written cookbook published in the United States contains authentic recipes for colonial favorites—pumpkin pudding, winter squash pudding, spruce beer, Indian slapjacks, and more. Introductory Essay and Glossary of colonial cooking terms. 80pp. 5⅜ × 8½. 24710-4 Pa. $3.50

101 PUZZLES IN THOUGHT AND LOGIC, C. R. Wylie, Jr. Solve murders and robberies, find out which fishermen are liars, how a blind man could possibly identify a color—purely by your own reasoning! 107pp. 5⅜ × 8½. 20367-0 Pa. $2.00

THE BOOK OF WORLD-FAMOUS MUSIC—CLASSICAL, POPULAR AND FOLK, James J. Fuld. Revised and enlarged republication of landmark work in musico-bibliography. Full information about nearly 1,000 songs and compositions including first lines of music and lyrics. New supplement. Index. 800pp. 5⅜ × 8¼. 24857-7 Pa. $14.95

ANTHROPOLOGY AND MODERN LIFE, Franz Boas. Great anthropologist's classic treatise on race and culture. Introduction by Ruth Bunzel. Only inexpensive paperback edition. 255pp. 5⅜ × 8½. 25245-0 Pa. $5.95

THE TALE OF PETER RABBIT, Beatrix Potter. The inimitable Peter's terrifying adventure in Mr. McGregor's garden, with all 27 wonderful, full-color Potter illustrations. 55pp. 4¼ × 5½. (Available in U.S. only) 22827-4 Pa. $1.75

THREE PROPHETIC SCIENCE FICTION NOVELS, H. G. Wells. *When the Sleeper Wakes, A Story of the Days to Come* and *The Time Machine* (full version). 335pp. 5⅜ × 8½. (Available in U.S. only) 20605-X Pa. $5.95

APICIUS COOKERY AND DINING IN IMPERIAL ROME, edited and translated by Joseph Dommers Vehling. Oldest known cookbook in existence offers readers a clear picture of what foods Romans ate, how they prepared them, etc. 49 illustrations. 301pp. 6⅛ × 9¼. 23563-7 Pa. $6.00

SHAKESPEARE LEXICON AND QUOTATION DICTIONARY, Alexander Schmidt. Full definitions, locations, shades of meaning of every word in plays and poems. More than 50,000 exact quotations. 1,485pp. 6½ × 9¼. 22726-X, 22727-8 Pa., Two-vol. set $27.90

THE WORLD'S GREAT SPEECHES, edited by Lewis Copeland and Lawrence W. Lamm. Vast collection of 278 speeches from Greeks to 1970. Powerful and effective models; unique look at history. 842pp. 5⅜ × 8½. 20468-5 Pa. $10.95

THE BLUE FAIRY BOOK, Andrew Lang. The first, most famous collection, with many familiar tales: Little Red Riding Hood, Aladdin and the Wonderful Lamp, Puss in Boots, Sleeping Beauty, Hansel and Gretel, Rumpelstiltskin; 37 in all. 138 illustrations. 390pp. 5⅜ × 8½. 21437-0 Pa. $5.95

THE STORY OF THE CHAMPIONS OF THE ROUND TABLE, Howard Pyle. Sir Launcelot, Sir Tristram and Sir Percival in spirited adventures of love and triumph retold in Pyle's inimitable style. 50 drawings, 31 full-page. xviii + 329pp. 6½ × 9¼. 21883-X Pa. $6.95

AUDUBON AND HIS JOURNALS, Maria Audubon. Unmatched two-volume portrait of the great artist, naturalist and author contains his journals, an excellent biography by his granddaughter, expert annotations by the noted ornithologist, Dr. Elliott Coues, and 37 superb illustrations. Total of 1,200pp. 5⅜ × 8.
Vol. I 25143-8 Pa. $8.95
Vol. II 25144-6 Pa. $8.95

GREAT DINOSAUR HUNTERS AND THEIR DISCOVERIES, Edwin H. Colbert. Fascinating, lavishly illustrated chronicle of dinosaur research, 1820's to 1960. Achievements of Cope, Marsh, Brown, Buckland, Mantell, Huxley, many others. 384pp. 5¼ × 8¼. 24701-5 Pa. $6.95

THE TASTEMAKERS, Russell Lynes. Informal, illustrated social history of American taste 1850's–1950's. First popularized categories Highbrow, Lowbrow, Middlebrow. 129 illustrations. New (1979) afterword. 384pp. 6 × 9.
23993-4 Pa. $6.95

DOUBLE CROSS PURPOSES, Ronald A. Knox. A treasure hunt in the Scottish Highlands, an old map, unidentified corpse, surprise discoveries keep reader guessing in this cleverly intricate tale of financial skullduggery. 2 black-and-white maps. 320pp. 5⅜ × 8½. (Available in U.S. only) 25032-6 Pa. $5.95

AUTHENTIC VICTORIAN DECORATION AND ORNAMENTATION IN FULL COLOR: 46 Plates from "Studies in Design," Christopher Dresser. Superb full-color lithographs reproduced from rare original portfolio of a major Victorian designer. 48pp. 9¼ × 12¼. 25083-0 Pa. $7.95

PRIMITIVE ART, Franz Boas. Remains the best text ever prepared on subject, thoroughly discussing Indian, African, Asian, Australian, and, especially, Northern American primitive art. Over 950 illustrations show ceramics, masks, totem poles, weapons, textiles, paintings, much more. 376pp. 5⅜ × 8. 20025-6 Pa. $6.95

SIDELIGHTS ON RELATIVITY, Albert Einstein. Unabridged republication of two lectures delivered by the great physicist in 1920–21. *Ether and Relativity* and *Geometry and Experience*. Elegant ideas in non-mathematical form, accessible to intelligent layman. vi + 56pp. 5⅜ × 8½. 24511-X Pa. $2.95

THE WIT AND HUMOR OF OSCAR WILDE, edited by Alvin Redman. More than 1,000 ripostes, paradoxes, wisecracks: Work is the curse of the drinking classes, I can resist everything except temptation, etc. 258pp. 5⅜ × 8½. 20602-5 Pa. $3.95

ADVENTURES WITH A MICROSCOPE, Richard Headstrom. 59 adventures with clothing fibers, protozoa, ferns and lichens, roots and leaves, much more. 142 illustrations. 232pp. 5⅜ × 8½. 23471-1 Pa. $3.95

PLANTS OF THE BIBLE, Harold N. Moldenke and Alma L. Moldenke. Standard reference to all 230 plants mentioned in Scriptures. Latin name, biblical reference, uses, modern identity, much more. Unsurpassed encyclopedic resource for scholars, botanists, nature lovers, students of Bible. Bibliography. Indexes. 123 black-and-white illustrations. 384pp. 6 × 9. 25069-5 Pa. $8.95

FAMOUS AMERICAN WOMEN: A Biographical Dictionary from Colonial Times to the Present, Robert McHenry, ed. From Pocahontas to Rosa Parks, 1,035 distinguished American women documented in separate biographical entries. Accurate, up-to-date data, numerous categories, spans 400 years. Indices. 493pp. 6½ × 9¼. 24523-3 Pa. $9.95

THE FABULOUS INTERIORS OF THE GREAT OCEAN LINERS IN HISTORIC PHOTOGRAPHS, William H. Miller, Jr. Some 200 superb photographs capture exquisite interiors of world's great "floating palaces"—1890's to 1980's: *Titanic, Ile de France, Queen Elizabeth, United States, Europa,* more. Approx. 200 black-and-white photographs. Captions. Text. Introduction. 160pp. 8⅞ × 11¾. 24756-2 Pa. $9.95

THE GREAT LUXURY LINERS, 1927–1954: A Photographic Record, William H. Miller, Jr. Nostalgic tribute to heyday of ocean liners. 186 photos of Ile de France, Normandie, Leviathan, Queen Elizabeth, United States, many others. Interior and exterior views. Introduction. Captions. 160pp. 9 × 12. 24056-8 Pa. $9.95

A NATURAL HISTORY OF THE DUCKS, John Charles Phillips. Great landmark of ornithology offers complete detailed coverage of nearly 200 species and subspecies of ducks: gadwall, sheldrake, merganser, pintail, many more. 74 full-color plates, 102 black-and-white. Bibliography. Total of 1,920pp. 8⅝ × 11¼. 25141-1, 25142-X Cloth. Two-vol. set $100.00

THE SEAWEED HANDBOOK: An Illustrated Guide to Seaweeds from North Carolina to Canada, Thomas F. Lee. Concise reference covers 78 species. Scientific and common names, habitat, distribution, more. Finding keys for easy identification. 224pp. 5⅜ × 8½. 25215-9 Pa. $5.95

THE TEN BOOKS OF ARCHITECTURE: The 1755 Leoni Edition, Leon Battista Alberti. Rare classic helped introduce the glories of ancient architecture to the Renaissance. 68 black-and-white plates. 336pp. 8⅜ × 11¼. 25239-6 Pa. $14.95

MISS MACKENZIE, Anthony Trollope. Minor masterpieces by Victorian master unmasks many truths about life in 19th-century England. First inexpensive edition in years. 392pp. 5⅜ × 8½. 25201-9 Pa. $7.95

THE RIME OF THE ANCIENT MARINER, Gustave Doré, Samuel Taylor Coleridge. Dramatic engravings considered by many to be his greatest work. The terrifying space of the open sea, the storms and whirlpools of an unknown ocean, the ice of Antarctica, more—all rendered in a powerful, chilling manner. Full text. 38 plates. 77pp. 9¼ × 12. 22305-1 Pa. $4.95

THE EXPEDITIONS OF ZEBULON MONTGOMERY PIKE, Zebulon Montgomery Pike. Fascinating first-hand accounts (1805–6) of exploration of Mississippi River, Indian wars, capture by Spanish dragoons, much more. 1,088pp. 5⅜ × 8½. 25254-X, 25255-8 Pa. Two-vol. set $23.90

A CONCISE HISTORY OF PHOTOGRAPHY: Third Revised Edition, Helmut Gernsheim. Best one-volume history—camera obscura, photochemistry, daguerreotypes, evolution of cameras, film, more. Also artistic aspects—landscape, portraits, fine art, etc. 281 black-and-white photographs. 26 in color. 176pp. 8⅜ × 11¼. 25128-4 Pa. $12.95

THE DORÉ BIBLE ILLUSTRATIONS, Gustave Doré. 241 detailed plates from the Bible: the Creation scenes, Adam and Eve, Flood, Babylon, battle sequences, life of Jesus, etc. Each plate is accompanied by the verses from the King James version of the Bible. 241pp. 9 × 12. 23004-X Pa. $8.95

HUGGER-MUGGER IN THE LOUVRE, Elliot Paul. Second Homer Evans mystery-comedy. Theft at the Louvre involves sleuth in hilarious, madcap caper. "A knockout."—Books. 336pp. 5⅜ × 8½. 25185-3 Pa. $5.95

FLATLAND, E. A. Abbott. Intriguing and enormously popular science-fiction classic explores the complexities of trying to survive as a two-dimensional being in a three-dimensional world. Amusingly illustrated by the author. 16 illustrations. 103pp. 5⅜ × 8½. 20001-9 Pa. $2.00

THE HISTORY OF THE LEWIS AND CLARK EXPEDITION, Meriwether Lewis and William Clark, edited by Elliott Coues. Classic edition of Lewis and Clark's day-by-day journals that later became the basis for U.S. claims to Oregon and the West. Accurate and invaluable geographical, botanical, biological, meteorological and anthropological material. Total of 1,508pp. 5⅜ × 8½.
21268-8, 21269-6, 21270-X Pa. Three-vol. set $25.50

LANGUAGE, TRUTH AND LOGIC, Alfred J. Ayer. Famous, clear introduction to Vienna, Cambridge schools of Logical Positivism. Role of philosophy, elimination of metaphysics, nature of analysis, etc. 160pp. 5⅜ × 8½. (Available in U.S. and Canada only) 20010-8 Pa. $2.95

MATHEMATICS FOR THE NONMATHEMATICIAN, Morris Kline. Detailed, college-level treatment of mathematics in cultural and historical context, with numerous exercises. For liberal arts students. Preface. Recommended Reading Lists. Tables. Index. Numerous black-and-white figures. xvi + 641pp. 5⅜ × 8½.
24823-2 Pa. $11.95

28 SCIENCE FICTION STORIES, H. G. Wells. Novels, *Star Begotten* and *Men Like Gods,* plus 26 short stories: "Empire of the Ants," "A Story of the Stone Age," "The Stolen Bacillus," "In the Abyss," etc. 915pp. 5⅜ × 8½. (Available in U.S. only)
20265-8 Cloth. $10.95

HANDBOOK OF PICTORIAL SYMBOLS, Rudolph Modley. 3,250 signs and symbols, many systems in full; official or heavy commercial use. Arranged by subject. Most in Pictorial Archive series. 143pp. 8⅛ × 11. 23357-X Pa. $5.95

INCIDENTS OF TRAVEL IN YUCATAN, John L. Stephens. Classic (1843) exploration of jungles of Yucatan, looking for evidences of Maya civilization. Travel adventures, Mexican and Indian culture, etc. Total of 669pp. 5⅜ × 8½.
20926-1, 20927-X Pa., Two-vol. set $9.90

DEGAS: An Intimate Portrait, Ambroise Vollard. Charming, anecdotal memoir by famous art dealer of one of the greatest 19th-century French painters. 14 black-and-white illustrations. Introduction by Harold L. Van Doren. 96pp. 5⅜ × 8½.
25131-4 Pa. $3.95

PERSONAL NARRATIVE OF A PILGRIMAGE TO ALMANDINAH AND MECCAH, Richard Burton. Great travel classic by remarkably colorful personality. Burton, disguised as a Moroccan, visited sacred shrines of Islam, narrowly escaping death. 47 illustrations. 959pp. 5⅜ × 8½. 21217-3, 21218-1 Pa., Two-vol. set $17.90

PHRASE AND WORD ORIGINS, A. H. Holt. Entertaining, reliable, modern study of more than 1,200 colorful words, phrases, origins and histories. Much unexpected information. 254pp. 5⅜ × 8½. 20758-7 Pa. $4.95

THE RED THUMB MARK, R. Austin Freeman. In this first Dr. Thorndyke case, the great scientific detective draws fascinating conclusions from the nature of a single fingerprint. Exciting story, authentic science. 320pp. 5⅜ × 8½. (Available in U.S. only) 25210-8 Pa. $5.95

AN EGYPTIAN HIEROGLYPHIC DICTIONARY, E. A. Wallis Budge. Monumental work containing about 25,000 words or terms that occur in texts ranging from 3000 B.C. to 600 A.D. Each entry consists of a transliteration of the word, the word in hieroglyphs, and the meaning in English. 1,314pp. 6⅞ × 10.
23615-3, 23616-1 Pa., Two-vol. set $27.90

THE COMPLEAT STRATEGYST: Being a Primer on the Theory of Games of Strategy, J. D. Williams. Highly entertaining classic describes, with many illustrated examples, how to select best strategies in conflict situations. Prefaces. Appendices. xvi + 268pp. 5⅜ × 8½. 25101-2 Pa. $5.95

THE ROAD TO OZ, L. Frank Baum. Dorothy meets the Shaggy Man, little Button-Bright and the Rainbow's beautiful daughter in this delightful trip to the magical Land of Oz. 272pp. 5⅜ × 8. 25208-6 Pa. $4.95

POINT AND LINE TO PLANE, Wassily Kandinsky. Seminal exposition of role of point, line, other elements in non-objective painting. Essential to understanding 20th-century art. 127 illustrations. 192pp. 6½ × 9¼. 23808-3 Pa. $4.50

LADY ANNA, Anthony Trollope. Moving chronicle of Countess Lovel's bitter struggle to win for herself and daughter Anna their rightful rank and fortune—perhaps at cost of sanity itself. 384pp. 5⅜ × 8½. 24669-8 Pa. $6.95

EGYPTIAN MAGIC, E. A. Wallis Budge. Sums up all that is known about magic in Ancient Egypt: the role of magic in controlling the gods, powerful amulets that warded off evil spirits, scarabs of immortality, use of wax images, formulas and spells, the secret name, much more. 253pp. 5⅜ × 8½. 22681-6 Pa. $4.00

THE DANCE OF SIVA, Ananda Coomaraswamy. Preeminent authority unfolds the vast metaphysic of India: the revelation of her art, conception of the universe, social organization, etc. 27 reproductions of art masterpieces. 192pp. 5⅜ × 8½.
24817-8 Pa. $5.95

CHRISTMAS CUSTOMS AND TRADITIONS, Clement A. Miles. Origin, evolution, significance of religious, secular practices. Caroling, gifts, yule logs, much more. Full, scholarly yet fascinating; non-sectarian. 400pp. 5⅜ × 8½.
23354-5 Pa. $6.50

THE HUMAN FIGURE IN MOTION, Eadweard Muybridge. More than 4,500 stopped-action photos, in action series, showing undraped men, women, children jumping, lying down, throwing, sitting, wrestling, carrying, etc. 390pp. 7⅞ × 10⅝.
20204-6 Cloth. $19.95

THE MAN WHO WAS THURSDAY, Gilbert Keith Chesterton. Witty, fast-paced novel about a club of anarchists in turn-of-the-century London. Brilliant social, religious, philosophical speculations. 128pp. 5⅜ × 8½. 25121-7 Pa. $3.95

A CEZANNE SKETCHBOOK: Figures, Portraits, Landscapes and Still Lifes, Paul Cezanne. Great artist experiments with tonal effects, light, mass, other qualities in over 100 drawings. A revealing view of developing master painter, precursor of Cubism. 102 black-and-white illustrations. 144pp. 8¾ × 6⅜. 24790-2 Pa. $5.95

AN ENCYCLOPEDIA OF BATTLES: Accounts of Over 1,560 Battles from 1479 B.C. to the Present, David Eggenberger. Presents essential details of every major battle in recorded history, from the first battle of Megiddo in 1479 B.C. to Grenada in 1984. List of Battle Maps. New Appendix covering the years 1967–1984. Index. 99 illustrations. 544pp. 6½ × 9¼. 24913-1 Pa. $14.95

AN ETYMOLOGICAL DICTIONARY OF MODERN ENGLISH, Ernest Weekley. Richest, fullest work, by foremost British lexicographer. Detailed word histories. Inexhaustible. Total of 856pp. 6½ × 9¼.
21873-2, 21874-0 Pa., Two-vol. set $17.00

WEBSTER'S AMERICAN MILITARY BIOGRAPHIES, edited by Robert McHenry. Over 1,000 figures who shaped 3 centuries of American military history. Detailed biographies of Nathan Hale, Douglas MacArthur, Mary Hallaren, others. Chronologies of engagements, more. Introduction. Addenda. 1,033 entries in alphabetical order. xi + 548pp. 6½ × 9¼. (Available in U.S. only)
24758-9 Pa. $11.95

LIFE IN ANCIENT EGYPT, Adolf Erman. Detailed older account, with much not in more recent books: domestic life, religion, magic, medicine, commerce, and whatever else needed for complete picture. Many illustrations. 597pp. 5⅜ × 8½.
22632-8 Pa. $8.50

HISTORIC COSTUME IN PICTURES, Braun & Schneider. Over 1,450 costumed figures shown, covering a wide variety of peoples: kings, emperors, nobles, priests, servants, soldiers, scholars, townsfolk, peasants, merchants, courtiers, cavaliers, and more. 256pp. 8⅜ × 11¼. 23150-X Pa. $7.95

THE NOTEBOOKS OF LEONARDO DA VINCI, edited by J. P. Richter. Extracts from manuscripts reveal great genius; on painting, sculpture, anatomy, sciences, geography, etc. Both Italian and English. 186 ms. pages reproduced, plus 500 additional drawings, including studies for *Last Supper, Sforza* monument, etc. 860pp. 7⅞ × 10¾. (Available in U.S. only) 22572-0, 22573-9 Pa., Two-vol. set $25.90

SIR HARRY HOTSPUR OF HUMBLETHWAITE, Anthony Trollope. Incisive, unconventional psychological study of a conflict between a wealthy baronet, his idealistic daughter, and their scapegrace cousin. The 1870 novel in its first inexpensive edition in years. 250pp. 5⅜ × 8½. 24953-0 Pa. $4.95

LASERS AND HOLOGRAPHY, Winston E. Kock. Sound introduction to burgeoning field, expanded (1981) for second edition. Wave patterns, coherence, lasers, diffraction, zone plates, properties of holograms, recent advances. 84 illustrations. 160pp. 5⅜ × 8¼. (Except in United Kingdom) 24041-X Pa. $3.50

INTRODUCTION TO ARTIFICIAL INTELLIGENCE: SECOND, ENLARGED EDITION, Philip C. Jackson, Jr. Comprehensive survey of artificial intelligence—the study of how machines (computers) can be made to act intelligently. Includes introductory and advanced material. Extensive notes updating the main text. 132 black-and-white illustrations. 512pp. 5⅜ × 8½. 24864-X Pa. $8.95

HISTORY OF INDIAN AND INDONESIAN ART, Ananda K. Coomaraswamy. Over 400 illustrations illuminate classic study of Indian art from earliest Harappa finds to early 20th century. Provides philosophical, religious and social insights. 304pp. 6⅝ × 9⅝. 25005-9 Pa. $8.95

THE GOLEM, Gustav Meyrink. Most famous supernatural novel in modern European literature, set in Ghetto of Old Prague around 1890. Compelling story of mystical experiences, strange transformations, profound terror. 13 black-and-white illustrations. 224pp. 5⅜ × 8½. (Available in U.S. only) 25025-3 Pa. $5.95

ARMADALE, Wilkie Collins. Third great mystery novel by the author of *The Woman in White* and *The Moonstone*. Original magazine version with 40 illustrations. 597pp. 5⅜ × 8½. 23429-0 Pa. $7.95

PICTORIAL ENCYCLOPEDIA OF HISTORIC ARCHITECTURAL PLANS, DETAILS AND ELEMENTS: With 1,880 Line Drawings of Arches, Domes, Doorways, Facades, Gables, Windows, etc., John Theodore Haneman. Sourcebook of inspiration for architects, designers, others. Bibliography. Captions. 141pp. 9 × 12. 24605-1 Pa. $6.95

BENCHLEY LOST AND FOUND, Robert Benchley. Finest humor from early 30's, about pet peeves, child psychologists, post office and others. Mostly unavailable elsewhere. 73 illustrations by Peter Arno and others. 183pp. 5⅜ × 8½. 22410-4 Pa. $3.95

ERTÉ GRAPHICS, Erté. Collection of striking color graphics: *Seasons, Alphabet, Numerals, Aces* and *Precious Stones*. 50 plates, including 4 on covers. 48pp. 9⅜ × 12¼. 23580-7 Pa. $6.95

THE JOURNAL OF HENRY D. THOREAU, edited by Bradford Torrey, F. H. Allen. Complete reprinting of 14 volumes, 1837–61, over two million words; the sourcebooks for *Walden*, etc. Definitive. All original sketches, plus 75 photographs. 1,804pp. 8½ × 12¼. 20312-3, 20313-1 Cloth., Two-vol. set $80.00

CASTLES: THEIR CONSTRUCTION AND HISTORY, Sidney Toy. Traces castle development from ancient roots. Nearly 200 photographs and drawings illustrate moats, keeps, baileys, many other features. Caernarvon, Dover Castles, Hadrian's Wall, Tower of London, dozens more. 256pp. 5⅜ × 8¼. 24898-4 Pa. $5.95